NON-ERGODIC GROUND MOTION MODELS FOR SITE SPECIFIC SEISMIC HAZARD ASSESSMENT AT NUCLEAR INSTALLATION SITES

The following States are Members of the International Atomic Energy Agency:

AFGHANISTAN
ALBANIA
ALGERIA
ANGOLA
ANTIGUA AND BARBUDA
ARGENTINA
ARMENIA
AUSTRALIA
AUSTRIA
AZERBAIJAN
BAHAMAS, THE
BAHRAIN
BANGLADESH
BARBADOS
BELARUS
BELGIUM
BELIZE
BENIN
BOLIVIA, PLURINATIONAL STATE OF
BOSNIA AND HERZEGOVINA
BOTSWANA
BRAZIL
BRUNEI DARUSSALAM
BULGARIA
BURKINA FASO
BURUNDI
CABO VERDE
CAMBODIA
CAMEROON
CANADA
CENTRAL AFRICAN REPUBLIC
CHAD
CHILE
CHINA
COLOMBIA
COMOROS
CONGO
COOK ISLANDS
COSTA RICA
CÔTE D'IVOIRE
CROATIA
CUBA
CYPRUS
CZECH REPUBLIC
DEMOCRATIC REPUBLIC OF THE CONGO
DENMARK
DJIBOUTI
DOMINICA
DOMINICAN REPUBLIC
ECUADOR
EGYPT
EL SALVADOR
ERITREA
ESTONIA
ESWATINI
ETHIOPIA
FIJI
FINLAND
FRANCE
GABON
GAMBIA, THE
GEORGIA
GERMANY
GHANA
GREECE
GRENADA
GUATEMALA
GUINEA
GUYANA
HAITI
HOLY SEE
HONDURAS
HUNGARY
ICELAND
INDIA
INDONESIA
IRAN, ISLAMIC REPUBLIC OF
IRAQ
IRELAND
ISRAEL
ITALY
JAMAICA
JAPAN
JORDAN
KAZAKHSTAN
KENYA
KOREA, REPUBLIC OF
KUWAIT
KYRGYZSTAN
LAO PEOPLE'S DEMOCRATIC REPUBLIC
LATVIA
LEBANON
LESOTHO
LIBERIA
LIBYA
LIECHTENSTEIN
LITHUANIA
LUXEMBOURG
MADAGASCAR
MALAWI
MALAYSIA
MALI
MALTA
MARSHALL ISLANDS
MAURITANIA
MAURITIUS
MEXICO
MONACO
MONGOLIA
MONTENEGRO
MOROCCO
MOZAMBIQUE
MYANMAR
NAMIBIA
NEPAL
NETHERLANDS, KINGDOM OF THE
NEW ZEALAND
NICARAGUA
NIGER
NIGERIA
NORTH MACEDONIA
NORWAY
OMAN
PAKISTAN
PALAU
PANAMA
PAPUA NEW GUINEA
PARAGUAY
PERU
PHILIPPINES
POLAND
PORTUGAL
QATAR
REPUBLIC OF MOLDOVA
ROMANIA
RUSSIAN FEDERATION
RWANDA
SAINT KITTS AND NEVIS
SAINT LUCIA
SAINT VINCENT AND THE GRENADINES
SAMOA
SAN MARINO
SAUDI ARABIA
SENEGAL
SERBIA
SEYCHELLES
SIERRA LEONE
SINGAPORE
SLOVAKIA
SLOVENIA
SOMALIA
SOUTH AFRICA
SPAIN
SRI LANKA
SUDAN
SWEDEN
SWITZERLAND
SYRIAN ARAB REPUBLIC
TAJIKISTAN
THAILAND
TOGO
TONGA
TRINIDAD AND TOBAGO
TUNISIA
TÜRKİYE
TURKMENISTAN
UGANDA
UKRAINE
UNITED ARAB EMIRATES
UNITED KINGDOM OF GREAT BRITAIN AND NORTHERN IRELAND
UNITED REPUBLIC OF TANZANIA
UNITED STATES OF AMERICA
URUGUAY
UZBEKISTAN
VANUATU
VENEZUELA, BOLIVARIAN REPUBLIC OF
VIET NAM
YEMEN
ZAMBIA
ZIMBABWE

The Agency's Statute was approved on 23 October 1956 by the Conference on the Statute of the IAEA held at United Nations Headquarters, New York; it entered into force on 29 July 1957. The Headquarters of the Agency are situated in Vienna. Its principal objective is "to accelerate and enlarge the contribution of atomic energy to peace, health and prosperity throughout the world".

IAEA-TECDOC-2093

NON-ERGODIC GROUND MOTION MODELS FOR SITE SPECIFIC SEISMIC HAZARD ASSESSMENT AT NUCLEAR INSTALLATION SITES

INTERNATIONAL ATOMIC ENERGY AGENCY
VIENNA, 2025

For further information on this publication, please contact:

External Events Safety Section
International Atomic Energy Agency
Vienna International Centre
PO Box 100
1400 Vienna, Austria
Email: Official.Mail@iaea.org

Printed by the IAEA in Austria
June 2025
https://doi.org/10.61092/iaea.9f3l-atpr

IAEA Library Cataloguing in Publication Data

Names: International Atomic Energy Agency.
Title: Non-ergodic ground motion models for site specific seismic hazard assessment at nuclear installation sites / International Atomic Energy Agency.
Description: Vienna : International Atomic Energy Agency, 2025. | Series: IAEA TECDOC series, ISSN 1011-4289 ; no. 2093 | Includes bibliographical references.
Identifiers: IAEAL 25-01761 | ISBN 978-92-0-110125-9 (paperback : alk. paper) | ISBN 978-92-0-110025-2 (pdf)
Subjects: LCSH: Nuclear facilities — Safety measures. | Nuclear facilities — Earthquake Effects. | Earthquake hazard analysis. | Hazard mitigation.

FOREWORD

In 2022, IAEA Safety Standards Series No. SSG-9, Seismic Hazards in Site Evaluation for Nuclear Installations, was updated to incorporate changes in the state of the practice in seismic hazard assessment, particularly the novel approaches utilized in ground motion characterization, over the past decade. One of the key updates is the acceptance of non-ergodic ground motion prediction equations as a method to estimate vibratory ground motions. Development and implementation of non-ergodic ground motion prediction equations in seismic hazard assessment is a relatively new practice in site evaluations for nuclear installations. This publication complements SSG-9 (Rev. 1) by describing the development and implementation methods for non-ergodic ground motion models as well as Member State experience.

The technical information and practical descriptions provided here will be valuable to nuclear power plant operators, regulatory bodies and technical support organizations in the field of seismic hazard assessment for new and existing nuclear installations.

The IAEA is grateful to all of those who contributed to the drafting and review of this publication, in particular N. Abrahamson (United States of America) and G. Senfaute (France). The IAEA officers responsible for this publication were A. Valentini and Z. Gulerce of the Division of Nuclear Installation Safety.

CONTENTS

1. INTRODUCTION

1.1.BACKGROUND

Requirement 16 of IAEA Safety Standards Series No. SSR-1, Site Evaluation for Nuclear Installations [1] states:

> **"An evaluation of ground motion hazards shall be conducted to provide the input needed for the seismic design or safety upgrading of the structures, systems and components of the nuclear installation, as well as the input for performing the deterministic and/or probabilistic safety analyses necessary during the lifetime of the nuclear installation."**

Paragraph 5.1 of IAEA Safety Standards Series No. SSG-9 (Rev. 1), Seismic Hazards in Site Evaluation for Nuclear Installations [2] recommends the use of empirical ground motion prediction equations as one of the methods to estimate vibratory ground motions in the evaluation of ground motion hazards for nuclear installations.

Global ground motion datasets (e.g. Refs [3]–[6]) combine empirical strong motion data from different regions within the same tectonic environment to provide ground motions from small-to-large magnitude earthquakes recorded at a certain range of distances. Typically, the ground motion models (GMMs)[1] that are developed by using these global datasets rely on the ergodic assumption and provide global and stable estimates of the median and its standard deviation to the user. As the size of the global strong ground motion datasets increases, the total standard deviations of the ergodic GMMs are stabilized around 0.5–0.75 (in natural log units). Global GMMs have been used in the seismic hazard assessment for nuclear installations very often, and this total standard deviation is implicitly included in the design basis ground motions. SSG-9 (Rev. 1) [2] underlines this significance by stating in para. 5.1 that "The variability associated with the prediction of vibratory ground motions from future earthquake is typically one of the largest sources of uncertainty*"*.

As the size of global ground motion datasets grow, repeatable (and systematic) characteristics of the empirical ground motion data become apparent. The GMMs are then modified to include these new effects, leading to more accurate GMMs with reduced aleatory variability. In the last decade, many global GMMs have incorporated region specific terms for broad regions or were regionalized by comparing their predictive performance with the local datasets. In addition, site specific median estimations and standard deviations were proposed by modelling the repeatable site effects, especially for nuclear installation sites, which led to a 10–15% reduction in the standard deviations and a change the median. These models can be considered as partially non-ergodic GMMs. Current practice of ground motion modelling is moving towards the fully

[1] Historically, empirical prediction of vibratory ground motion was estimated using 'attenuation equations'. However, the equation is not only relevant to the seismic wave attenuation but also includes the scaling of ground motions with source and site parameters. Therefore, scientific societies use the term "Ground Motion Prediction Equation (GMPE)" and SSG-9 (Rev. 1) [2] adopted the term GMPE instead of 'attenuation equation'. Current GMPEs include a large number of parameters to represent the non-ergodicity and many coefficients in the equations are provided as tables, hence the scientific societies use the term "Ground Motion Model (GMM)". In this publication, the terms of GMPE and GMM are synonymous, and both are used in accordance with the background of the paragraphs.

non-ergodic approach, in which the systematic and repeatable site, source, and path effects are explicitly modelled. Using fully non-ergodic GMMs will result in a significant reduction in the standard deviation and may result in a shift in the median, which will eventually lead to relevant changes in the design basis ground motions. Because the shift in the median can be positive or negative, using non-ergodic GMMs does not necessarily imply a systematic reduction in the hazard estimates. The non-ergodic hazard estimates can be either larger or smaller than the ergodic hazard estimates.

Paragraph 5.10 of SSG-9 (Rev. 1) [2] expresses the possibility of using non-ergodic GMMs for nuclear installations, by stating that "if non-ergodic GMPEs are to be used, all coefficients should be properly identified to represent the ground motions for specific conditions." A certain reduction in the standard deviation is also noted in para. 5.10 of SSG-9 (Rev. 1) [2], which states that "If ergodic GMPEs are to be used, they will generally be able to capture overall ground motion characteristics with fewer parameters, although the standard deviation might be larger than for non-ergodic GMPEs."

Even if the possible use of non-ergodic GMMs is acknowledged, detailed guidance on the development and implementation of partially or fully non-ergodic GMMs for the seismic hazard assessment of nuclear installations is not given in SSG-9 (Rev. 1) [2].

1.2. OBJECTIVE

The main objective of this publication is to complement SSG-9 (Rev. 1) [2] by describing the differences in ergodic, partially non-ergodic, and full non-ergodic ground motion models and by detailing the development and implementation of non-ergodic GMMs for the evaluation of seismic hazards in site evaluation of nuclear installations. In addition, this publication has collected the state-of-the-art knowledge and practices related to the application of non-ergodic GMMs in Member States, in probabilistic seismic hazard assessment (PSHA).

1.3. SCOPE

This publication addresses the non-ergodic approach in ground motion modelling. The methodology defined in this publication for the development of partially or full non-ergodic GMMs may be implemented in the evaluation of seismic hazards for new or existing nuclear power plants (NPPs) and other nuclear installations in any seismotectonic environment. It also covers the comprehensive information related to the data needed to develop partially and full non-ergodic GMMs and changes in the median estimations, epistemic uncertainty in the median, and total standard deviations of the non-ergodic models compared to the ergodic models. The scope does not include the ergodic GMMs that are already addressed in other IAEA publications (e.g. Ref. [7]) and in many other publications.

This publication is intended for use by regulatory bodies responsible for establishing regulatory requirements and operating and technical support organizations responsible for the evaluation of seismic hazards at a nuclear installation site.

1.4. STRUCTURE

Section 2 of this publication describes the terminology and concepts of ergodicity, partial non-ergodicity, and full non-ergodicity in ground motion modelling. Section 3 discusses the ground motion data that may be used for developing non-ergodic GMMs, including data from on-site seismic monitoring system and regional empirical ground motion datasets. Section 4 is devoted to the aleatory and epistemic uncertainties, their significance in the non-ergodic ground motion modelling, and possible reduction in the uncertainty in the non-ergodic approach. Section 5 presents the methodology to develop partially non-ergodic GMMs for the site term. Section 6 presents the methodology to develop fully non-ergodic GMMs. Section 7 discusses the support that may be provided by 3-D ground motion simulations to estimate the non-ergodic terms in GMMs. Section 8 presents the implementation of non-ergodic ground motion models in PSHA and a short summary of the evaluation of non-ergodic GMMs with independent observations. Section 9 summarizes the main conclusions.

Annexes of this publication include examples of the state of the art knowledge and practices related to the application of non-ergodic GMMs in Member States. Annex I is an example from the Czech Republic for the evaluation of attenuation anomalies and their consideration in the partially non-ergodic model for the Bohemian Massif. Annex II presents the Japanese regulations and practices concerning the non-ergodic ground motion models. Annex III presents the probabilistic seismic hazards assessment of Krsko 2 NPP in Slovenia using a fully non-ergodic GMM.

2. ERGODICITY FOR GROUND MOTION MODELS

This section provides the state-of-the-practice for ergodic and non-ergodic ground motion models. The terminology used in this section to define the partially or fully non-ergodic ground motion models is used for the rest of this publication in a consistent manner.

2.1. GENERAL FRAMEWORK

Ergodicity of a process shows that every sequence or sizable sample is equally representative of the whole. To use a simple analogy to help understand the difference between ergodic and non-ergodic models used in seismic hazard assessment, it is possible to consider the distribution of the heights of males in a human population. In the ergodic model, the heights of males throughout the world are combined into a single distribution. Based on a large data set, the median height may be 175 cm with a standard deviation of 20 cm. In the ergodic approach for modelling the expected height of males in the future, this distribution is assumed to apply everywhere: for all regions of the world, the distribution of the heights has the same mean and standard deviation. Because this is based on a large sample of observations, the epistemic uncertainty in the mean height is small. In the non-ergodic model, the heights of males are measured for small regions. For example, consider a small island with a population of 25 males with a median height of 155 cm and a standard deviation of 10 cm. For the non-ergodic model of this distribution, there is a reduction in the standard deviation (from the ergodic sigma of 20 cm to the non-ergodic sigma of 10 cm), and there is a shift in the median by -20 cm from 175 cm to 155 cm. However, due to the small number of observations for the non-ergodic model, the epistemic uncertainty in the shift from the ergodic to non-ergodic median value is non-zero.

Another analogy is the use of a broad diffused seismicity zone with uniform seismicity distribution versus a gridded seismicity distribution in seismic source characterization (Fig. 1). The uniform seismicity distribution shown in Fig. 1(a) is the analogue for the ergodic model: all independent earthquakes inside the areal source zone are combined to get an average rate of earthquakes, and this average rate is assumed to apply to all sites in the zone. The gridded seismicity distribution given in Fig. 1(b) is the analogue for the non-ergodic model: the rate of earthquakes is estimated for each grid using a smoothing kernel. There are fewer data to determine the rate of earthquakes in individual grids than for the entire zone. Therefore, the epistemic uncertainty in the rate of earthquakes in each grid is larger than the epistemic uncertainty for the average rate of earthquakes within the large uniform zone with more observations.

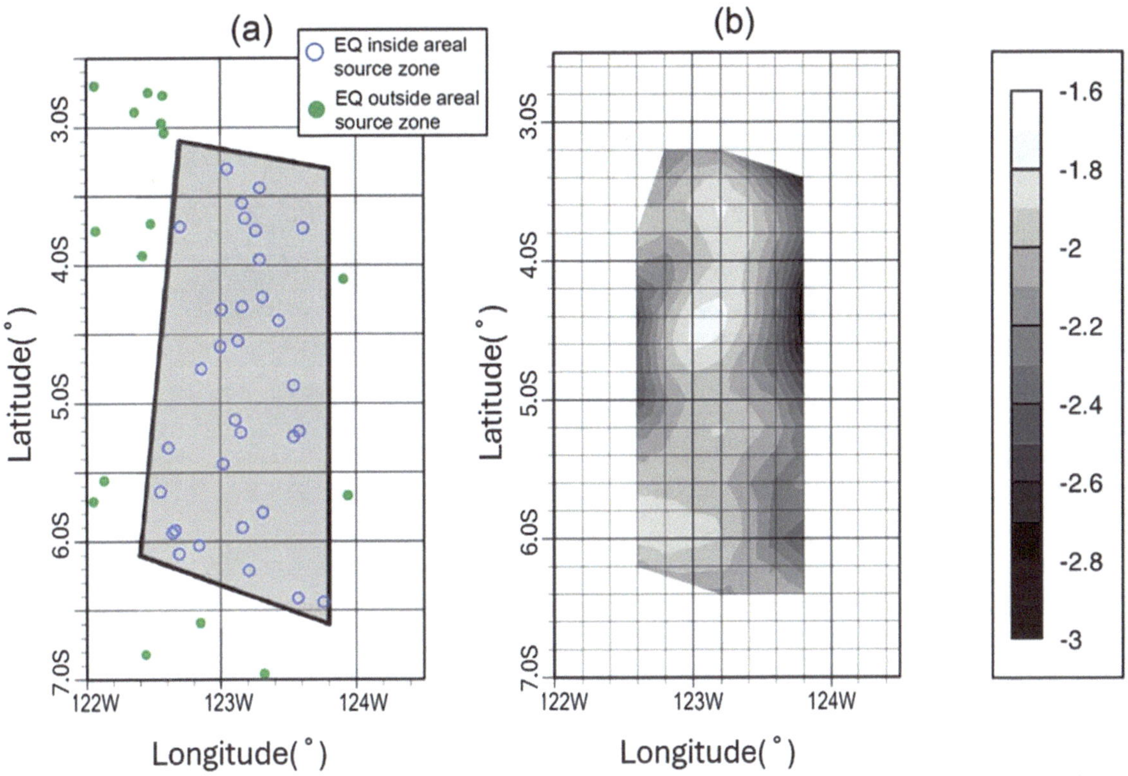

FIG. 1. (a) The diffused seismicity zone in which the earthquakes are combined into a constant average rate (analogue for ergodic model), (b) the gridded seismicity zone with the rate of earthquakes estimated for each grid (analogue for the non-ergodic model). Gray scale shows the variation in the rate of events in logarithmic units. Horizontal and vertical axes are the coordinates in longitude and latitude in degrees, respectively.

2.2.ERGODIC GROUND MOTION MODELS

Traditionally, GMMs are constructed by combining empirical strong motion data from different regions of the world with similar tectonic regimes: active crustal regions, stable continental regions, and subduction zones. The advantage of combining data is that it leads to a data set with better coverage of the magnitude-distance space, which improves the robustness of the fit of the model to the data; however, there are regional differences in the source, distance and site amplification scaling of strong motions within the same broad tectonic category. For example, the path scaling of GMMs includes a combination of geometrical spreading (the logarithmic distance scaling term in GMMs) and anelastic attenuation (the linear distance scaling term), which depends on the 3-D shear wave velocity and attenuation characteristics of the crustal structure. There can be significant differences in the velocity structure and attenuation characteristics between regions (e.g. Japan versus for California) and even within broad regions (e.g. southern California versus northern California). A simple comparison of recorded strong motions of a well recorded earthquake with model predictions reveals these differences within a region as shown in Fig. 2. As demonstrated by Ref. [8], the ergodic GMM with a fixed anelastic term (black curve) overpredicted the observed data, while the non-ergodic GMM proposed by Ref. [9] has predicted the recorded data better by allowing for the local adjustment of the anelastic attenuation term (red curve). Using an ergodic approach, there would be a large negative event term (difference between the black and red curves at a distance of about 50 km) but using the non-ergodic approach with a steeper distance slope, the event term would be the difference between the black and red curves at 1 km distance, which is a smaller shift than for the ergodic GMM.

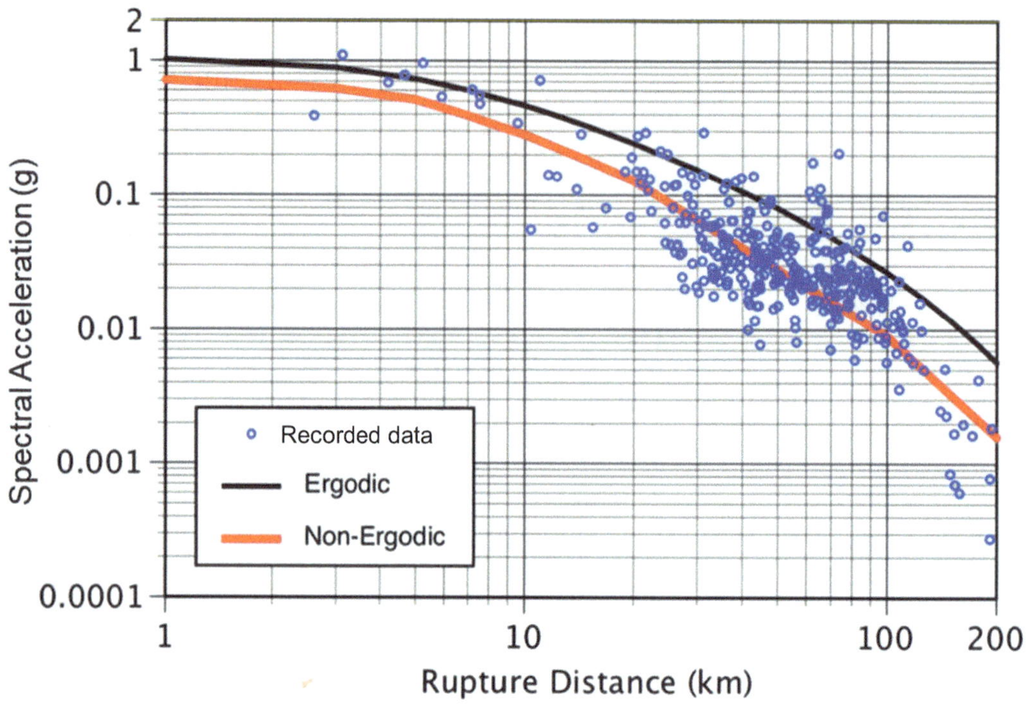

FIG. 2. Regional differences in the distance attenuation from the 2014 M=6.0 South Napa, California earthquake. Blue points are the spectral accelerations at T=0.2 second, normalized to the same site conditions. Black and red curves are the median predictions of an ergodic and non-ergodic GMM (adapted from [10]).

The traditional approach of combining recordings from different broad regions into a single global GMM is called the 'ergodic approach in ground motion modelling', and it typically neglects these regional differences. An ergodic process is a random process, in which the distribution of a random variable in space is the same as the distribution of that random variable at a single point when sampled over time. An ergodic assumption is made when the seismic hazard assessment treats the spatial differences in the ground motions at specific sites as random variability that can occur at any site over time [11]. Making the ergodic assumption (by using an ergodic GMM) allows trading the variability in space over a short time for variability over all time at a single location.

The ergodic assumption has been widely used in the development of GMMs; however, it results in a large total ergodic aleatory standard deviation because the repeatable site, source, and path effects are ignored. Larger standard deviations can have a large impact on hazard estimates, in particular for critical facilities that are designed for ground motions with long return periods [12]. A more accurate median ground motion with smaller aleatory variability could be achieved by relaxing the ergodic assumption and including non-ergodic terms in the model. In a non-ergodic seismic hazard assessment, site specific source, path, and site terms are included in the non-ergodic GMM, and there is a corresponding reduction in the aleatory standard deviation due to the improved fit to the observed data [13].

Compared to the ergodic GMM, the aleatory standard deviation of a non-ergodic GMM is always reduced, but the change in the median can be either positive or negative depending on the site and source pair. The hazard estimate will be affected by both the change in the median and the reduction in the aleatory standard deviation. As an example, the probability distribution functions (PDF) of the ground motions using an ergodic GMM and a non-ergodic GMM are shown in Fig. 3(a). The three values of the shift in the median are for the 5th, 50th, and 95th fractiles for a site without any site specific ground motion data to constrain the non-ergodic terms. The non-ergodic GMM has a narrower distribution which leads to steeper hazard curves when compared to the ergodic hazard curve (purple and black lines in Fig. 3(b)). The different estimates of the change in the median shift the hazard curve along the x-axis. As shown in Fig. 3(b), including the non-ergodic GMM can result in either a decrease or an increase in the hazard depending on value of the shift in the median. As the ground motion data are collected to constrain the non-ergodic GMM, the epistemic range of the non-ergodic terms will be reduced, but the average shift in the median could be positive or negative.

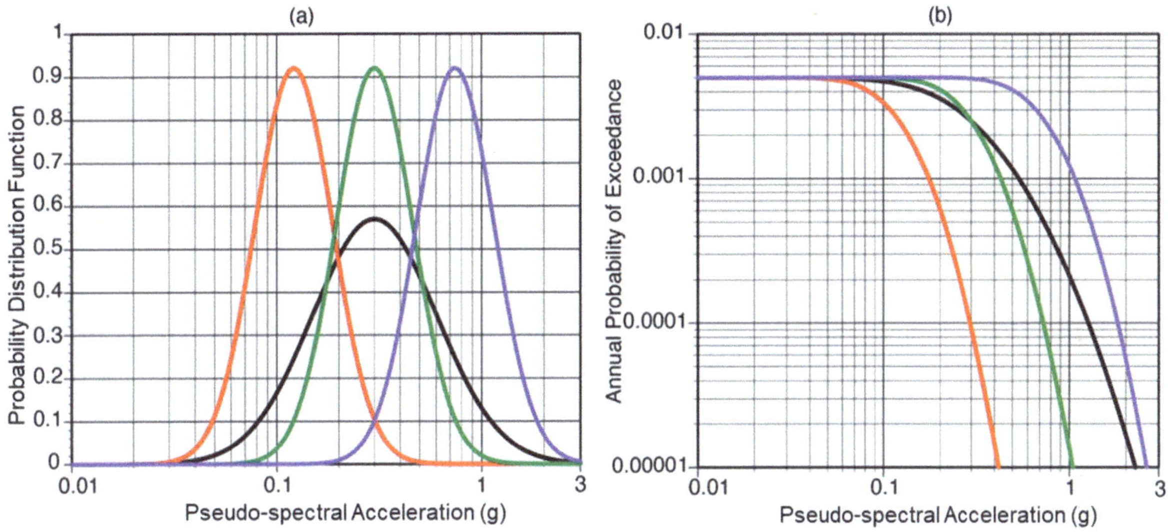

FIG. 3. (a) Probability distribution function of ground motions for ergodic and non-ergodic models (5th, 50th, and 95th fractiles with red, green, and purple curves, respectively). (b) Hazard curves for a single source (M7) with a recurrence interval of 200 years that implement the ground motion models given in (a).

2.3.PARTIALLY TO FULLY NON- ERGODIC GROUND MOTION MODELS

The ergodic GMMs combine data from all regions in the world for the same broad tectonic category (Fig. 4, top), while non-ergodic GMMs are region- or site specific. There are different components of non-ergodic GMMs: (1) regional terms, (2) site terms, (3) path terms, and (4) source terms (Fig. 4, bottom). GMMs that include all of these terms are called fully non-ergodic GMMs, whereas GMMs that include only some of these non-ergodic terms are called partially non-ergodic models. The non-ergodic terms in the GMMs are used to adjust only the linear (amplitude-independent) scaling of the GMMs. This allows for the use of large data sets of ground motions from small magnitude earthquakes to constrain the non-ergodic terms while still be applicable to the large magnitude events that often control the hazard.

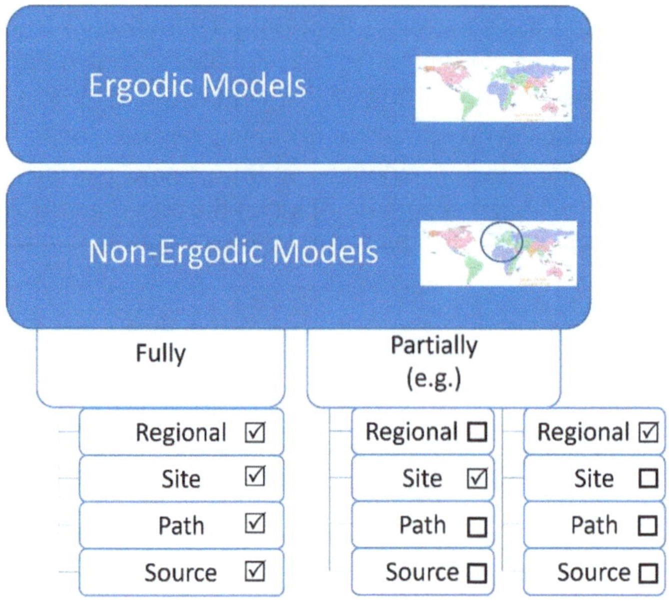

FIG. 4. Ergodic and non-ergodic GMMs. In this example, only the non-ergodic site terms or a regional model are included, but any combination is possible for a partially non-ergodic model.

2.3.1. Partially non-ergodic models with non-ergodicity in the regional terms

Typically, the first step in developing non-ergodic GMMs is to develop a regionalized GMM by modifying some of the coefficients of a global ergodic GMM to account for systematic differences between broad regions. This can be done during the model development phase by including region specific terms or can be done after the model development by adding region specific adjustment terms to the ergodic GMM based on the regional subset of strong motion data. The GMMs with region specific terms are partially non-ergodic GMMs. They use the ergodic assumption within a broad region, but they account for regional differences in the distance scaling, site scaling, and constant terms. Examples of regionalized GMMs include the NGA-West2 GMMs [14] and the NGA-Subduction GMMs [15] that have region specific terms for the linear distance scaling, the site amplification scaling, and a regional constant. Figure 5(a) shows that the median predictions of an NGA-Subduction GMM (proposed by Ref. [16]) vary significantly by region for a magnitude 8 subduction interface event at a distance of 50 km for rock site conditions.

In most regionalized GMMs, the source scaling (magnitude and depth scaling) is not regionalized as often there is not adequate data within a specific region to constrain the source scaling terms; however, some studies have found a clear need to regionalize the magnitude scaling, particularly for smaller magnitudes. For example, the small magnitude ($M<6$) scaling in the independent Turkish strong motion data set is clearly different from the global scaling. To account for this systematic difference, a regional adjustment factor was added to the magnitude scaling in Ref. [17] as shown in Fig. 5(b) for a magnitude 5.5 event. This may reflect a difference in the estimation of magnitudes of small-to-moderate earthquakes for different regions, or it could represent a regional difference in the stress drop scaling with magnitude.

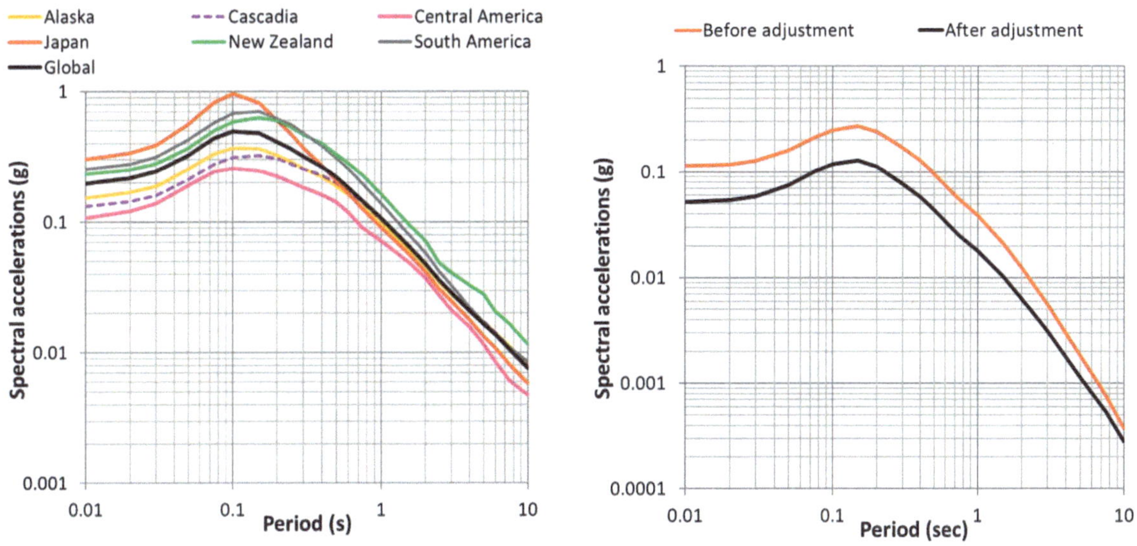

(a) (b)

FIG. 5. Examples of differences in the median predictions of regionalized GMMs for different regions. (a) subduction interface earthquakes; (b) crustal earthquakes.

The aleatory standard deviation for regionalized GMMs is often based on the combined data set (i.e. an ergodic assumption is made for the value of the standard deviation), but in some cases, regional differences in the standard deviation are used. The NGA-West2 GMMs [14] are examples of regionalized models with a single aleatory standard deviation model for all regions. The NGA-Subduction GMM proposed by Ref. [16] is an example of a regionalized model with different standard deviations for different regions (Fig. 6).

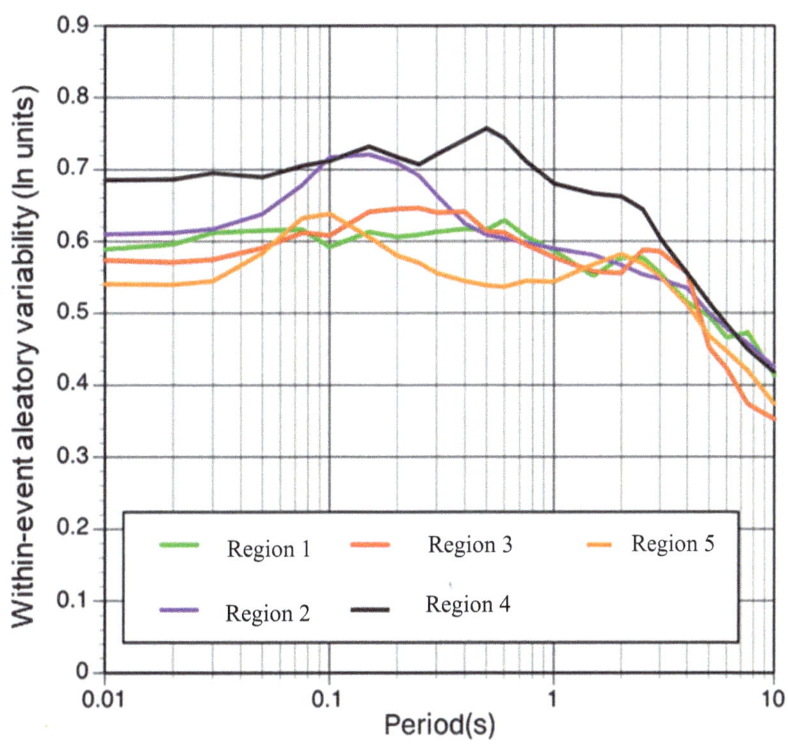

FIG. 6. Example of differences in the within-event aleatory variability for subduction earthquakes for different regions (adapted from [16]).

2.3.2. Partially non-ergodic models with non-ergodicity in the site terms

A second form of a partially non-ergodic GMM modifies the median and aleatory standard deviation of ergodic GMMs to account for a site specific site term, also known as the 'single-station sigma' approach as described in Ref. [7]. In this approach, the magnitude and distance scaling are represented by the scaling implemented in the ergodic GMM (in other words, are not modified); however, the average site amplification at the target site is treated as a site specific term that is repeatable and systematic. A constant site specific adjustment term is added to the median of the ergodic or regionalized GMM.

For the partially non-ergodic GMMs with non-ergodicity in the site term, the within-event residual is separated into the site specific adjustment term and the remaining within-site residual. By including the non-ergodic site term, the aleatory standard deviation is reduced to the single-station sigma value. Typically, the single-station sigma is 10–25% smaller than the ergodic sigma. The amount of the reduction depends on how well the site terms in the GMM work. For regions in which time-averaged shear wave velocity at the top 30 m (V_{S30}) works well as a proxy for site amplification (e.g. California, New Zealand, Türkiye), the reduction will be smaller because there is less to adjust. In regions in which V_{S30} does not work well as a proxy for site amplification (e.g. Japan, South America), the reduction will be larger because there will be a larger range of the site specific site terms.

2.3.3. Partially non-ergodic models with non-ergodicity in the path terms

Partially non-ergodic GMMs have been extended to include spatial variation in the distance scaling. Typically, the non-ergodic site term is also included when a non-ergodic path term is included, but if there are insufficient recordings per station to constrain the non-ergodic site terms, the path term can be estimated without the site term. The estimation of the non-ergodic path terms is more complicated than the non-ergodic site terms because the path terms depend on both the site and source locations.

An example of path terms based on the 3-D simulations for a site in the Los Angeles region is shown in Fig. 7 [18]. Figure 7(a) shows in map view the paths of different source locations for a specific site. The distance attenuation depends strongly on the azimuth to the source. The path effects can be large, here they range from -0.6 to 0.6 natural logarithmic (ln) units (a factor of 0.55–1.8). Figure 7(b) compares the distance attenuation for the ergodic GMM (blue), the partially non-ergodic GMM with only site terms (black), and the non-ergodic GMM with site and path terms (red). For the site term, the adjustment to the ergodic GMM is a constant shift (from blue curve to black curve) as this non-ergodic site term applies to all source locations (all distances). In contrast, the non-ergodic path terms can result in a larger or smaller ground motions than the ergodic GMM depending on the location of the source (azimuth and distance from the site). The jagged nature of the distance scaling of the non-ergodic GMM indicates that there are regions of focusing and defocusing of the seismic waves. The non-ergodic GMM has better resolution for specific site/source pairs, but since the non-ergodic GMMs include more model terms, the epistemic uncertainty is often larger for non-ergodic GMMs.

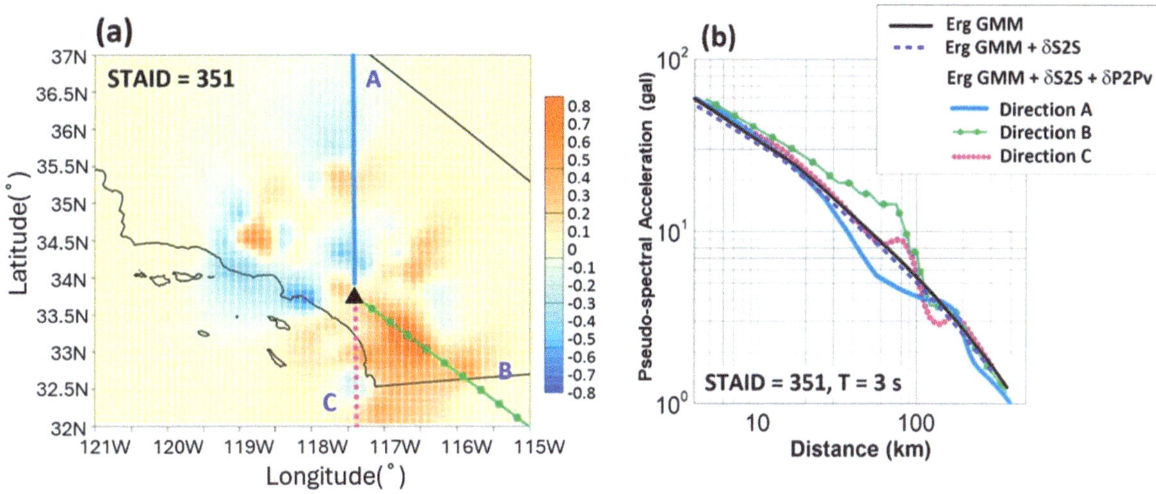

FIG. 7. Example of the path effects for spectral acceleration at T=3 seconds from 3-D simulations. (b) The distance scaling of the spectral acceleration at T=3 seconds in the three directions shown in (a) for the ergodic GMM, the ergodic GMM with the non-ergodic site term, and the ergodic GMM with both the non-ergodic site term and path term. Scale bar shows the change in the ground motion in natural log units (adapted from [18]). Horizontal and vertical axes in (a) are the coordinates in longitude and latitude in degrees, respectively.

2.3.4. Partially non-ergodic ground motion models with non-ergodicity in the source terms

The non-ergodic source term represents the difference between the median source term for a specific earthquake location and the median source term in the GMM. This is a constant shift in the amplitude of the median ground motion. This non-ergodic source term may capture differences in the median stress drop of earthquakes in a particular sub-region compared to the median stress drop for the whole region that is in the ergodic GMM. As an example, the median non-ergodic source terms for California are shown in Fig. 8 [19]. For this case, the non-ergodic source terms range from -0.3 to 0.3 ln units (a factor of 0.75–1.25) which is a smaller range that seen for the site and path effects. The main source of data for the source terms are small magnitude events because they are more frequent and cover a wider spatial range than the large magnitude events. It is not known if the non-ergodic source terms estimated from small magnitude events are applicable to large magnitude events. When the data shows regional differences in the median stress drops for small magnitude events, it may imply that there will be a similar regional difference in the median stress drops for large magnitude events.

In the estimation of the non-ergodic terms, regional differences in the source terms from small magnitude earthquakes need to be included in the model to avoid mapping source differences into site or path effects. However, for the application into seismic hazard assessment, the non-ergodic source terms can be either included or excluded depending on the assumption of the applicability of the regional source terms from small magnitude events to large magnitude events. Since the between-event standard deviation is usually much smaller than the within-event standard deviation, the effect of the non-ergodic source term is usually smaller than the effect of the non-ergodic site and path terms.

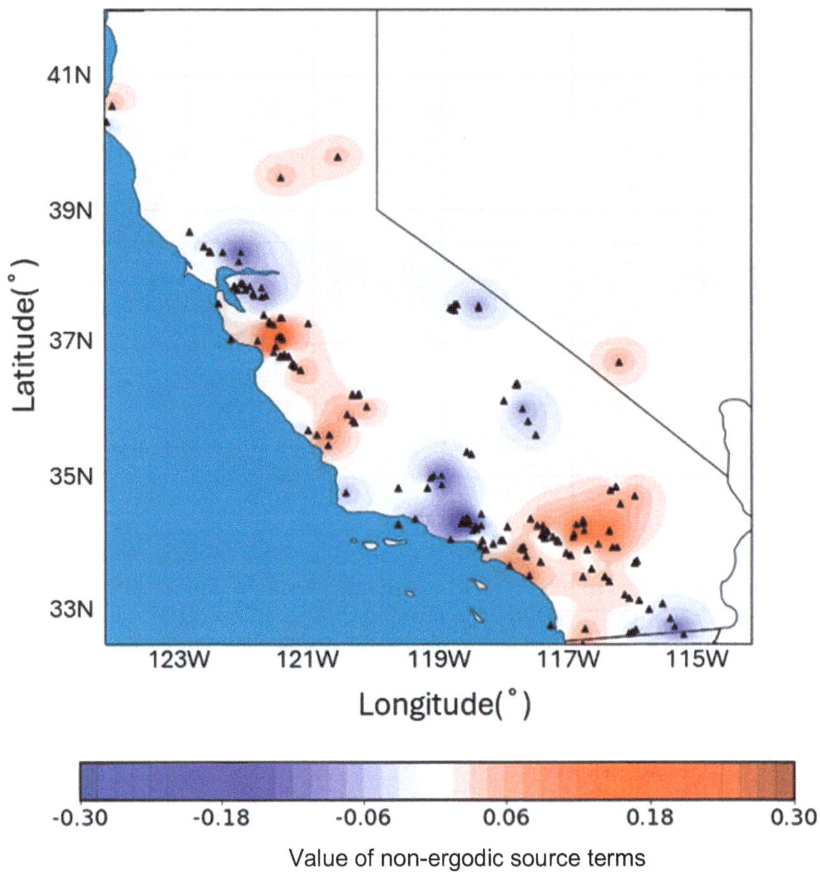

FIG. 8. Example of non-ergodic source terms for California. Black triangles show the locations of the earthquakes (adapted from [19]). Horizontal and vertical axes are the coordinates in longitude and latitude in degrees, respectively.

2.3.5. Fully non-ergodic ground motion models

The fully non-ergodic GMM includes three changes from the ergodic GMM: (1) the regional aleatory standard deviation is reduced, (2) there is a change (either increase or decrease) in the median, and (3) the epistemic uncertainty in the change in the median is included as an additional node in the GMM logic tree. The aleatory variability can either be modelled by the regional value or by a site specific value (see Section 5 for details).

As an example of the reduction in the aleatory variability, the ratio of the non-ergodic to ergodic standard deviations estimated for two different regions is shown in Fig. 9. The ratio ranges from 0.4 to 0.6, corresponding to 60–80% of the variance of the ergodic GMM due to systematic site, path, and source effects. The reduction is larger for one of the regions because the site response is more variable, so there is more opportunity to reduce the ergodic standard deviation by including the non-ergodic site terms.

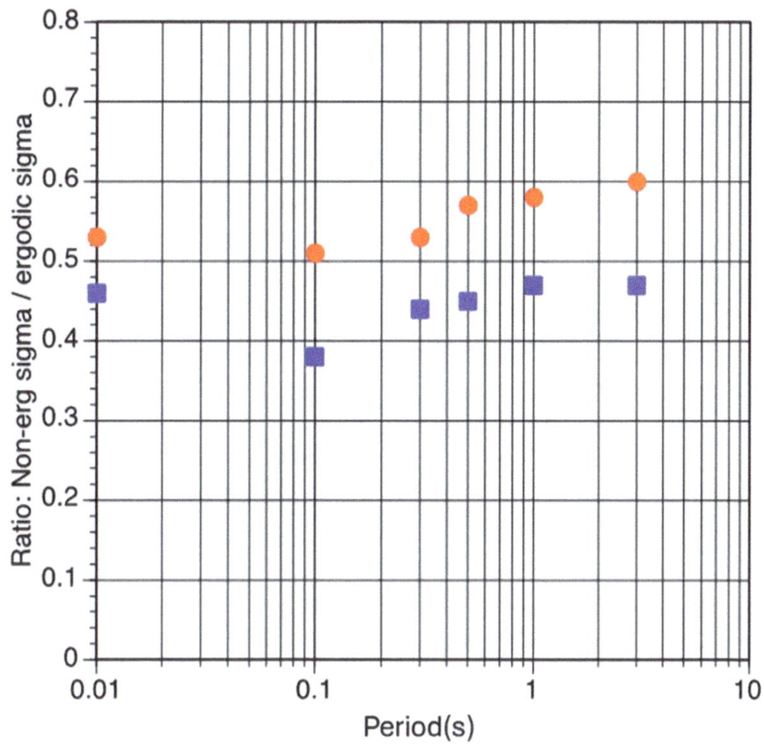

FIG. 9. Examples of the reduction of the standard deviation using a fully non-ergodic GMM compared to the ergodic GMM for two different active crustal regions (red: region1, blue: region 2).

The median of an ergodic GMM is modified by the consideration of specific adjustments to the site, path, and source terms for developing a fully non-ergodic GMM. The change in the median will vary by the site and source location. The standard deviation of the change in the median over the entire region depends on the reduction in the aleatory variability. If the ratio of the non-ergodic sigma to the ergodic sigma is 0.5, then the standard deviation of the change in the median ground motion over the region will be 0.87 of the ergodic standard deviation.

To justify the use of the reduced aleatory variability in non-ergodic GMMs, the epistemic uncertainty in the non-ergodic terms needs to be included in the ground motion logic tree. If there is no local data available to constrain the non-ergodic terms, then the epistemic uncertainty will be large. The classification of uncertainty into aleatory variability and epistemic uncertainty depends on the level of model simplification. If the non-ergodic approach is used for a site without data to constrain the non-ergodic terms, aleatory variability associated with the non-ergodic effects in the ergodic GMM will be shifted to epistemic uncertainty in the non-ergodic GMM. As local data are collected or site specific simulations are generated, then the central estimate of the median changes, and the epistemic uncertainty range is also reduced.

3. DATA REQUIRED TO DEVELOP NON-ERGODIC GROUND MOTION MODELS

Site specific seismic hazard assessment rely mainly on regional and site specific data. Given the increasing size of modern national and regional seismic networks over the last decades, more data has become available to develop non-ergodic models. Furthermore, at some nuclear installations, permanent site specific networks have been deployed. Nevertheless, the availability of a large amount of regional (e.g. less than 300 km around the site), local (up to 30 km), and site specific data is not common. This lack of data prevents the derivation of statistically robust site specific GMMs for nuclear installations.

This section is intended to describe the data and associated metadata necessary to develop non-ergodic ground motion models. This overview is limited to the description of the data needed. Data to be used for such developments needs to be stored and available in well structured databases, enabling easy access for model development. To cover these aspects, a separate technical document that focuses on the best practices for deployment of seismic instrumentation and gives recommendations on the collection, processing and analyses of the ground motions recorded at the site and in the area surrounding the nuclear installation site is currently under preparation. This new technical document will also include an extensive presentation of a ground motion database structure suitable for use in the development of non-ergodic models.

In addition to collecting ground motion recordings, it is essential to have high quality site characterization at ground motion recording stations. This characterization includes the knowledge of the shallow velocity profile, deep velocity profile, and geotechnical parameters for the soil layers. The shallow velocity profile can be characterized by V_{S30}, and the deep velocity profile may be characterized by the depth to V_S=1 km/s ($Z_{1.0}$) or V_S=2.5 km/s ($Z_{2.5}$) parameters among others. In addition, information on the 2-D or 3-D nature of the velocity profiles in the site region is important for the non-ergodic path and site terms. The methods for developing 3-D velocity profiles are discussed in Annex II. Development of well constrained non-ergodic GMMs depends on having appropriate knowledge of the instrumentation of the recording stations and having site conditions that are well known and characterized.

This section also distinguishes between regional data that is useful to develop source, and path non-ergodic adjustments and site data that is used to constrain and refine the site term and to develop site response analysis. Several considerations on the required data to develop ergodic ground motion models (as discussed in Ref. [7]) also apply to non-ergodic models. The focus of this section are the additional needs in terms of data availability and qualification for use in the development of non-ergodic ground motion models. Other independent data such as macro-seismic intensities, which may be used to evaluate non-ergodic ground motion models, is addressed in Section 8.

3.1. GENERAL CONSIDERATIONS ON DATA QUALITY

The quality and quantity of ground motion recordings available to develop non-ergodic ground motion models may vary greatly, depending on the development of national, regional, or local seismic networks and on the level of seismicity. Despite these differences, datasets used to develop non-ergodic ground motion models need to consider the following criteria:

— Uniform processing of the recordings is important, following modern standards adopted in the development of international databases (e.g. Refs [3]–[6]) and as described in Refs [20], [21];
— Such processing will clearly assess the usable frequency band of each record, based on the cut-off frequencies of the filter applied to remove the noise from the seismic signal. Band-pass frequencies are preferably determined based on a record specific analysis of the signal-to-noise ratio (SNR) rather than on fixed frequencies based on general considerations (e.g. Ref. [22]). This would allow maximizing the usable frequency band avoiding the contamination of the ground motions by noise. Because the development of non-ergodic GMMs requires a large amount of data, almost inevitably, the collected data set will contain a large number of small magnitude recordings (down to magnitude 3 and possibly lower). The characterization of the SNR is thus a particularly important step to establish the usable frequency range for such records;
— Information about the seismic instrumentation system needs to be clearly provided to allow for the verification that the instrument type is fit for purpose and that the data has been appropriately corrected for the instrument response;
— Information on the location and installation conditions of the instrument will also be considered and recordings from free-field instruments will be used in the analyses. Recordings from instruments located within buildings, dams or other structures could be used after having verified that the structure response does not affect the recordings in the frequency range of interest for the analyses;
— The study needs to demonstrate that all the available data are collected for the development of the non-ergodic ground motion model.

A well structured database is accompanied by a parametric table (a so called 'flat file') summarizing all the relevant metadata concerning the events, the recording stations, and the waveforms for each datum. Such a parametric table might include response spectral and Fourier Amplitude Spectral (FAS) ordinates for each record.

3.2. REGIONAL DATA

The regional dataset has two components: (i) the recorded ground motions, and (ii) associated metadata.

3.2.1. Recorded ground motions

Ground motions recorded at multiple sites from multiple sources at a regional scale may be used to derive non-ergodic GMMs. Ideally, recordings with sufficient spatial coverage around the target site in terms of both distance and azimuth ranges are collected and the ground motion dataset samples multiple sources and ray paths relevant for the hazard assessment. For target sites located close to a national border, the ground motion records from the neighbouring seismological networks may also be collected and treated with a homogeneous processing scheme. The availability of data at smaller scales (near regional, site vicinity and site area: see para. 3.7 of SSG-9 (Rev. 1) [2]) will help to constrain non-ergodic path and site terms.

Figure 10 shows the spatial distribution of the earthquakes and the recording stations in Central Chile used for the development of non-ergodic GMM [23]. Multiple recordings at the same station enable the estimation of systematic site effects. Similarly, multiple recordings from the same earthquake provide insights into the estimation of systematic source effects. Repeatable wave propagation paths between a given pair of earthquake and station locations can reveal systematic path effects. Another example is shown in Fig. 11, considering the data used to develop a non-ergodic GMM for California [24]. The station density is not homogeneous as there are areas with much more data than others. The lack of data in some regions does not prevent the development of non-ergodic GMM, but it will be reflected in larger epistemic uncertainties of the non-ergodic terms for such areas of low coverage.

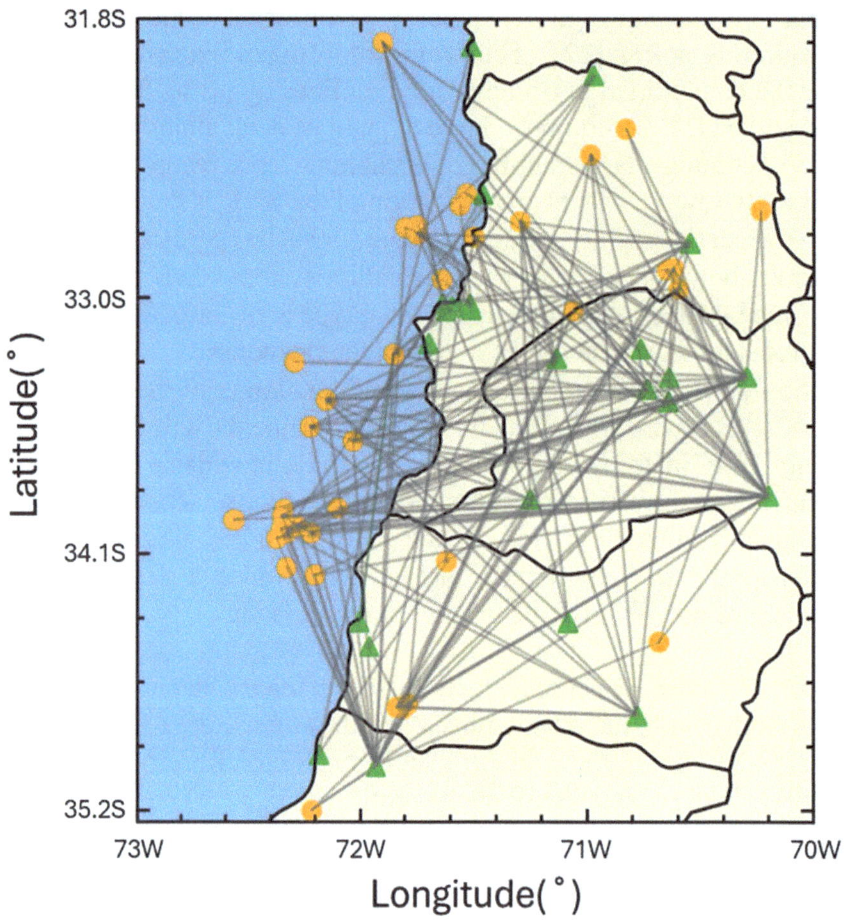

FIG. 10. Example of the distribution of earthquake epicentres (orange points), stations (green triangles), and wave propagation paths (grey lines) in Central Chile (adapted from [23]). Horizontal and vertical axes are the coordinates in longitude and latitude in degrees, respectively.

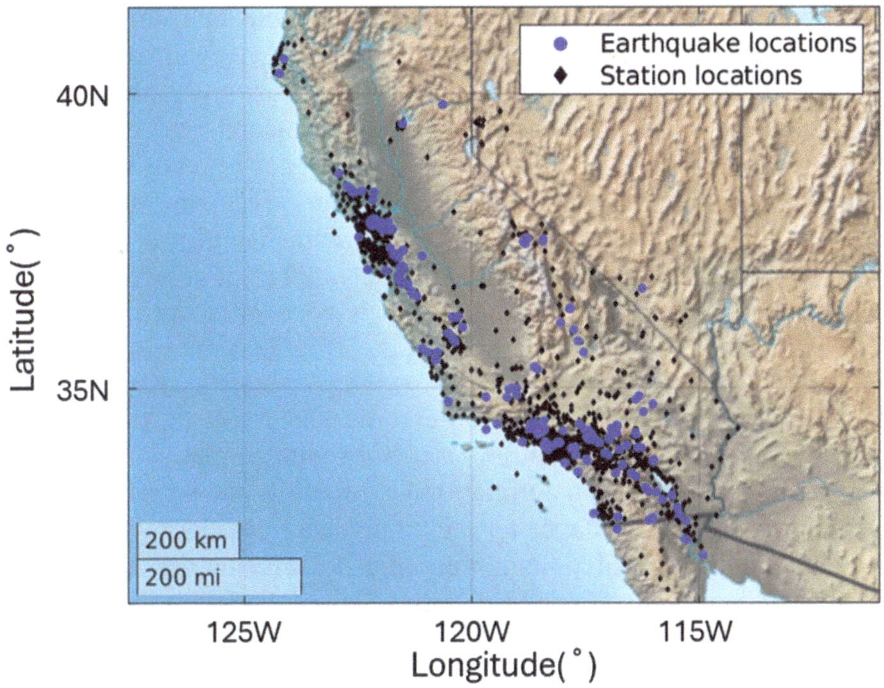

FIG. 11. Earthquakes' epicentres (stars) and stations (triangles) used by Ref. [24] to develop the non-ergodic GMM for California (adapted from [24]). Horizontal and vertical axes are the coordinates in longitude and latitude in degrees, respectively.

As an example of magnitude-distance distributions in two regions characterized by significantly different level of seismicity and data quantity, Fig. 12 compares the available recordings in California [19] and France [25]. As expected, for the case of California, small to moderate events (magnitudes below 5) represent the major contribution to the dataset, but recordings for larger events are still available. On the other hand, in France, the available data exclusively come from small-to-moderate magnitude events. Nonetheless, thanks to the collection of small magnitude recordings (2< M< 5), non-ergodic GMM can still be developed, as shown by Ref. [25]. To have as much data as possible, significant effort needs to be devoted to the collection and processing of recordings from small magnitude earthquakes. Collecting data that are not typically available publicly will likely require direct interaction with data providers or national seismic network management.

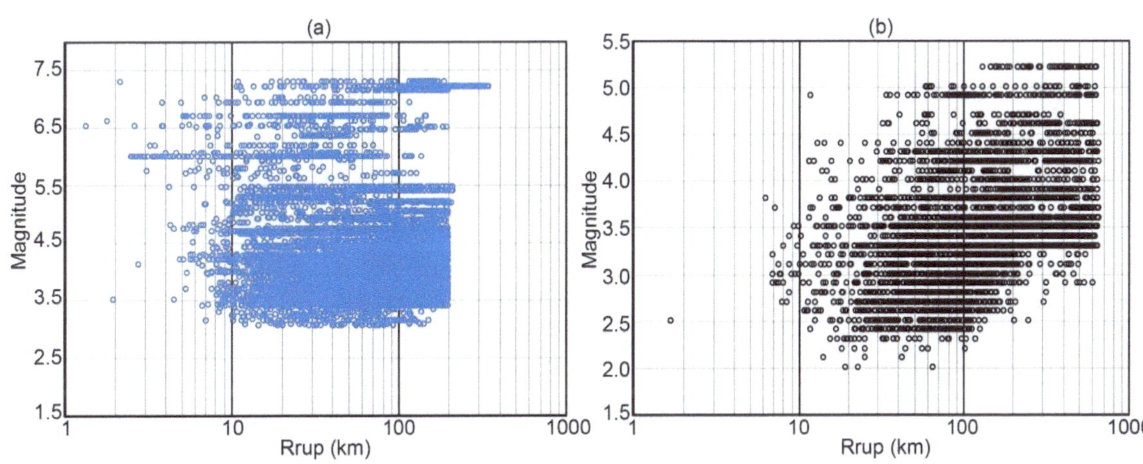

FIG. 12. Magnitude-distance distribution of recordings available for the development of non-ergodic ground motion models in California (left, Ref. [19]) and in France (right, Ref. [25]). Horizontal axis denoted by R_{RUP} shows the closest distance of the site to the rupture plane.

The evaluation of the path term requires observed ground motions that sample a range of ray paths to compute linear distance scaling adjustments for all the relevant source-to-site paths considered in the hazard calculations. To have robust estimates of the path terms, at least 10 recordings are sufficient to sample the same path unit element (e.g. cell). Similarly, the calculation of the non-ergodic source and site terms requires enough recordings per event and site, respectively (i.e. > 10) to have reliable estimates of mean adjustments and relatively small epistemic uncertainties. The suggested numbers of minimum recordings are indicative and non-ergodic terms could still be developed with smaller number of recordings. In this case, careful consideration is needed as limited knowledge will be reflected in larger epistemic uncertainties associated to the non-ergodic terms.

If no recordings are available at the target site, non-ergodic site terms can still be approximated based on recordings collected at several instrumented sites with similar geology located close to the target site, if available. The spatial correlation length of the site terms for the region can be used to determine the maximum distance for including data from nearby recording stations. For cases without data at the target site, the non-ergodic site terms at the target site will have large epistemic uncertainties, as discussed in detail in Section 5. Similarly, source terms for event locations where few data are available, or no events occurred could be evaluated based on events are larger distances from the target source if the spatial correlation length is large enough. Details regarding correlation lengths are given in Section 6. Thus, it is important to not only collect the data at the site area scale, but to encompass all the available stations at the near regional and site vicinity scales to provide constraints on the non-ergodic term and epistemic uncertainties.

3.2.2. Metadata

The calculation of non-ergodic terms is based on residuals of observed ground motions with respect to an ergodic GMM, an accurate source (e.g. moment magnitude - M_W, style-of-faulting) and recording site metadata (e.g. V_{S30}) is necessary to develop a non-ergodic GMM. Therefore, a robust site characterization for recording stations is vitally important. Information from independent studies conducted in the region can be used to retrieve information on the attenuation, source, and site terms. A non-exhaustive list of useful metadata is given below:

- M_W estimates, particularly for small magnitude earthquakes;
- Precise earthquake location and depth estimates;
- Shear wave velocity profile at instrumented stations;
- V_{S30} estimates at instrumented stations (based on geotechnical or geophysical data or proxies);
- Estimates of basin depth or sediment thickness ($Z_{1.0}$ or $Z_{2.5}$ parameter);
- Estimates of fundamental frequency of soil column at instrumented sites;
- Estimates of fault-mechanism for the considered events;
- Estimates of stress parameter for the considered events;
- Information on the attenuation structure and on the 3-D velocity structure in the region.

The calculation of the path terms will be less sensitive to the quality of the source and site metadata such as M_W, V_{S30} and other parameters used in the reference GMM since they are based on the decomposition of the within-site residuals where systematic source and site effects have been removed. It is also worth noting that if full site characterization for every strong motion recording station is not available, an increase in the epistemic uncertainty is expected.

3.3. SITE DATA

In some applications, collection of a regional ground motion dataset may not be feasible due, for example, to the lack of in-place seismic networks and/or to the very low seismic activity of the region. In such cases, if the nuclear installation site is instrumented and site specific data are available, partially non-ergodic GMM with the adjustment to the site term may be developed. Even in low to moderately active areas, a few years at most is often enough to obtain sufficient recordings to estimate linear site terms.

The use of site recordings to compute site specific adjustments for the site term and related uncertainties have been discussed by several authors (e.g. Refs [26]–[28]). Reference [28] presents the calculation of partially non-ergodic adjustments of the site term (δS2S) at seismic stations in Europe and the Middle East, and specifically discuss the estimation of epistemic uncertainties on δS2S as a function on the number of available recordings at the site. Analysis results show that the standard error of δS2S decreases rapidly with the increasing number of recordings, leading to a reduction of the standard error by more than 50% when 10 records are available (Fig. 13(a)). This is in good agreement with Ref. [26], suggesting that the epistemic uncertainties on δS2S can be evaluated by the standard deviation of δS2S divided by the square root of the number of recordings. Since the δS2S term is derived empirically, the recorded data needs to be within the magnitude-distance-site conditions range covered by the ergodic GMM and used for the calculation of residuals. For this reason, empirical GMMs are being extended to small magnitude events.

An example of non-ergodic site terms being developed from a small number of recordings at Diablo Canyon nuclear power plant site is provided in Fig. 13(b) [29]. In this case, there are only three recordings at the site, so the epistemic uncertainty in the non-ergodic site term is large; however, the non-ergodic site terms are non-zero.

Additional data and parameters derived from the earthquake time histories collected at the site or from geophysical and geotechnical site data can be used to inform and evaluate if the site term from ergodic GMM is applicable to the target site, if available. Among those are:

- Site amplification with respect to one or more reference stations, assessed using different techniques;
- Soil and seismic velocity profiles;
- Horizontal-to-vertical spectral ratios to determine the fundamental frequency of the site.

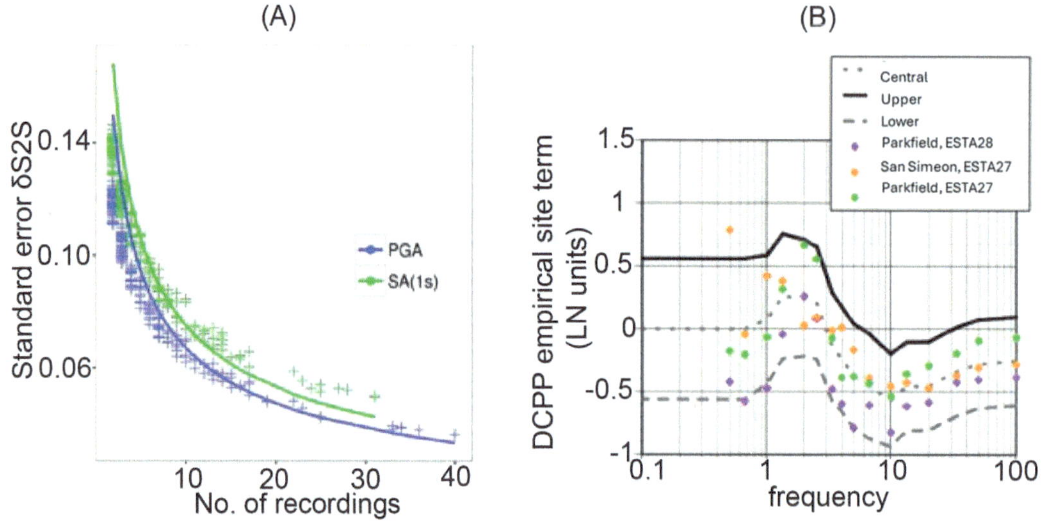

FIG. 13. (a) Standard error of δ_{S2S} as a function of number of recordings available for peak ground acceleration and spectral acceleration at 1 second (blue and green crosses, respectively) (adapted from [28]). (b) Non-ergodic site terms estimated from a small number of recordings at the Diablo Canyon nuclear power plant site (adapted from [29]).

3.4.SIMULATED GROUND MOTIONS

3-D numerical simulations have been used to estimate site and path terms to support the development of fully non-ergodic GMM. As ground motions may be obtained for any source region, site, and path with much higher density and sampling compared to empirical observations, simulations can be used to support the development of non-ergodic models. The methods use and incorporate ground motion simulations in non-ergodic GMM are discussed in Section 7. Here, the focus is on the minimum requirements on simulated data, describing the steps to qualify the simulated ground motions adopted to derive non-ergodic models. Important qualifications of simulated data for use in non-ergodic GMMs are as follows:

1. Validation of the simulation approach: The adopted approach needs to be based on 3-D wave propagation (e.g. Refs [30], [31]) to provide meaningful information of path and site terms. Many different simulation approaches are available, and it is important that the adopted method is carefully validated by comparing simulations with observations from previous earthquakes. Most of the 3-D simulation approaches provide long-period ground motions (e.g. spectral accelerations for periods generally larger than 1 or 2 seconds). Simplified models (e.g. the stochastic model from Ref. [32]) are then used to extend the frequency content of simulated ground motions to shorter periods. It is important that such a frequency limit is clearly identified to define the frequency range of interest. Finally, the ground motion simulation needs be performed for the target site region and not imported from other host regions based on some similarity criteria.

2. Validation of the 3-D crustal/soil model and underlying data: The 3-D crustal model adopted for the simulations needs to be scientifically sound and based on well established, published, and peer-reviewed data with clear documentation of the underlying data and the model development. Moreover, a validation of the 3-D model using observations (e.g. by simulating earthquakes in the region) is desirable. If

possible, the impact of alternative 3-D models on the simulated ground motion needs to be assessed and guide the characterization of epistemic uncertainties.

3. Adequate simulation set up: The simulations need to be performed for realistic earthquake scenarios in the magnitude-distance range of interest for the seismic hazard. Therefore, consideration of multiple source locations and realizations of the rupture process to sample several paths multiple times is necessary. Finally, the grid on receiver stations needs to be sufficiently dense (e.g. 1 km spacing) to allow for capturing spatial differences in the site terms.

3.5. SITES WITH LIMITED OR NO GROUND MOTION DATA

A common assumption is that non-ergodic GMMs can only be used at sites with sufficient data to constrain the non-ergodic terms, and the sites with limited or no data need to continue to use ergodic GMMs. As a result, the critical question is 'what is the minimum amount of data required to use the non-ergodic approach?'. However, this assumption and related question does not address the epistemic uncertainty.

A fundamental concept in seismic hazard analysis is that lack of data implies large epistemic uncertainty. The lack of data does not change the physical behavior of ground motions so that they become ergodic; rather, lack of data results in large uncertainty in the non-ergodic terms. If there is a lack of data, then simple approaches for estimating the median and epistemic uncertainty of the non-ergodic terms can be used. To be consistent with the epistemic uncertainty, the aleatory variability needs to be the reduced non-ergodic standard deviation.

This issue of how to treat regions with little or no local data is related to the 'process-based' versus 'model-based' treatment of aleatory variability and epistemic uncertainty as described in the next section. For the mean hazard, the results are not affected by the partition of aleatory and epistemic uncertainty. If there is no data, then the mean hazard computed using an ergodic GMM is equal to the mean hazard computed using non-ergodic GMMs with full epistemic uncertainty on the non-ergodic terms. If the goal of the PSHA is only to compute the mean hazard, then ergodic GMMs can be used for sites without data; however, if the goal of the PSHA is to estimate the fractiles of the epistemic uncertainty on the hazard in addition to the mean hazard, then non-ergodic GMMs with wide epistemic uncertainties need to be used.

This issue is also related to the evaluation of PSHA using observations. The observed ground motions will reflect the non-ergodic site, path, and source effects. If the observations are compared to hazard results based on ergodic GMMs which are not applicable to any specific site, then it is expected that the observations will be inconsistent with the ergodic hazard curves.

For application of the non-ergodic approach to regions with sparse data, the size of the epistemic uncertainty in the non-ergodic terms can be estimated using data from other regions. This makes an ergodic assumption about the variance of the non-ergodic terms and allows for site and source specific non-ergodic terms. That is, an ergodic assumption is made about the covariance structure, and the non-ergodic approach is still applied to sites without data. For regions without region specific data to constrain the non-ergodic terms, the mean hazard will be unchanged from the ergodic model, but the non-ergodic approach will show the limitations of the available data which is masked with the ergodic approach.

In the early adoption of the non-ergodic approach, the epistemic uncertainties in the resulting hazard will often be large. It is important to communicate the current limitations of our ability to constrain the seismic hazard and to consider those uncertainties when making engineering judgments about adequate seismic safety. In addition, the epistemic uncertainty in the hazard due to the uncertainty in the non-ergodic terms shows how much the estimate of the seismic hazard can be improved with the collection of additional empirical data or the generation of additional data using analytical modelling. As data are collected or simulated, the epistemic uncertainty will be reduced. This approach provides a framework for improving estimates over time as new data are collected.

4. ALEATORY AND EPISTEMIC UNCERTAINTIES IN NON-ERGODIC GROUND MOTION MODELS

Aleatory uncertainty is described as the 'uncertainty inherent in a phenomenon' and epistemic uncertainty is described as the 'uncertainty attributable to incomplete knowledge about a phenomenon, which affects the ability to model it' in Ref. [33]. The terms aleatory variability and epistemic uncertainty are used in PSHA to differentiate between the aspects of the seismic source characterization and ground motion characterization models that are treated as 'alternative and credible source and ground motion models' (epistemic uncertainty) and 'apparent randomness due to use of simplified models that is treated as randomness within each alternative source and ground motion model' (aleatory variability). These two definitions highlight that epistemic uncertainty can be either related to the physical data-generating process, or to the simplified model of the process [34]. The process-based epistemic uncertainty described by a set of hazard curves that are expected to bound the true hazard at the site, whereas the model-based epistemic uncertainty represents the uncertainty that the simplified model is the best model for the given the level of model simplification.

Standard practice in PSHA uses the model-based approach. In general, the epistemic fractiles from the model-based approach will not capture the range of the true hazard, and they are not intended to do so. The unmodelled systematic effects due to use of simplified models can have significant effects on the hazard at a site. By treating the unmodelled effects as aleatory variability, the range on the hazard curves due to systematic but unmodelled effects is collapsed into a hazard curve with a flatter slope.

In ground motion characterization, the unmodelled effects lead to a range of possible values of the ground motion parameter for a given scenario with simplified parameterization (e.g. magnitude, distance, and site condition). This range of ground motion values is described by the PDF of the GMM. Typically, the PDF of the ground motion is modelled by a lognormal distribution which is treated as aleatory variability. In the ground motion characterization model, typically more than one GMM is selected, and a weight is assigned to each GMM in the logic tree. These alternative GMMs represent the epistemic uncertainty in the range of alternative credible simplified GMMs.

The concepts of aleatory variability and epistemic uncertainty have been widely used in PSHA since the 1980s; however, the apparent non-uniqueness of the separation between aleatory variability and epistemic uncertainty has been often matter of concern because it gives the impression that such a separation is arbitrary, eventually putting in doubt its usefulness. With the model-based approach commonly used in PSHA, the aleatory variability is mainly composed of apparent randomness due to model simplification. There is also chaotic randomness for highly non-linear systems, and inherent randomness due to quantum effects, but these are small compared to the apparent randomness due to model simplification [34]. The amount of apparent randomness depends on the level of model simplification: with less model simplification, there are fewer unmodelled effects and less apparent randomness. As a result, the classification into epistemic uncertainty and aleatory variability is not absolute. However, once a model is selected with a certain level of simplification, there is no ambiguity in their separation. The non-ergodic GMMs are examples of different classification of aleatory variability and epistemic uncertainty depending on the level of simplification of the GMM. As more non-ergodic terms are added to the GMM (site, path, and source), fewer effects are left unmodelled and classified as aleatory variability, and more effects are treated as systematic effects and classified as epistemic uncertainty.

4.1. DEPENDENCE OF ALEATORY VARIABILITY AND EPISTEMIC UNCERTAINTY ON GROUND MOTION MODELLING

For an ergodic GMM, the ground motion (Y_{es}) at station *s* from earthquake *e* can be written as:

$$\ln(Y_{es}) = \mu_{erg}(M, R, S, \ldots) + \delta B_e + \delta W_{es} \tag{1}$$

in which $\mu_{erg}(M, R, S, \ldots)$ is the ergodic median prediction, δB_e and δW_{es} are the between-event and within-event residuals with standard deviations of τ and φ, respectively [7]. The total ergodic aleatory standard deviation (σ_{erg}) is given by Eq. (2):

$$\sigma_{erg} = \sqrt{\phi^2 + \tau^2} \tag{2}$$

The ground motion from a deterministic simulation with a given slip-distribution, hypocentre location, rupture velocity, source-time function and a given 3-D velocity structure (as used in physics-based PSHA [35]) from earthquake *e*, at site *s* can be written as:

$$\ln(Y_{es}) = \mu_{det}(M_e, site\ location, VS_{3D}, slip, hypo, \ldots) + \delta_{0,es} \tag{3}$$

in which $\delta_{0,es}$ is the residual relative to the deterministic simulation due to inherent randomness. The recorded ground motions will also include randomness due to chaotic effects and uncertainty at the quantum level (inherent randomness) that are not part of the deterministic simulation. Physics-based GMMs that sample the range of physical parameters as input into deterministic simulations can lead to the same amount of variability of the empirical ground motion data [36], indicating that the standard deviation of $\delta_{0,es}$ is small relative to the standard deviation from the variability of the inputs to the deterministic simulations. The $\delta_{0,es}$ term is included for completeness, but it is assumed to be negligible.

Following the definitions given in Ref. [37], there are three types of aleatory variability in GMMs:

- — Apparent randomness due to systematic source, path, or site effects that are practical to model, but which the GMM developer chose not to model;
- — Apparent randomness due to systematic source, path, or site effects that are not practical to model at this time;
- — Randomness due to chaotic highly non-linear behaviour and quantum level uncertainty that is impossible to model.

Using these three classes of aleatory variability, Ref. [37] describes the ground motion in terms of systemic source, path, and site effects that are practical to model, systematic effects that are not practical to model, and effects that are impossible to model as in Eq. (4):

$$\ln(Y_{es}) = [\mu_{erg}(M, R, S, \ldots)] + [\delta L2L_e + \delta P2P_{es} + \delta S2S_s] \tag{4}$$

$$+[\delta B_e^0 + \delta WSP_{es} + \delta AMP_s] + \delta_{0,es}$$

In Eq. (4), the first term include systematic effects that are included in the model, the second group of terms are the systematic source effects ($\delta L2L$), path effects ($\delta P2P$), and site effects ($\delta S2S$) that are practical to model, but which were not modelled, and the third group of terms are systematic source effects (δB_e^0), path effects (δWP_{es}), and site effects (δAMP_s) that are not practical to model. In GMMs, the δWSP_{es} and δAMP_s terms are combined into the within-site term. In an ergodic GMM, the first and second groups of terms are treated as aleatory variability that can occur for any site and source combination in the region.

Assuming that the residuals are independent, the aleatory standard deviation of the ergodic GMM is given by Eq. (5). The notations for the standard deviations are provided in Table 1.

$$\sigma_{erg} = \sqrt{\tau_{L2L}^2 + \phi_{P2P}^2 + \phi_{S2S}^2 + \tau_0^2 + \phi_{WP}^2 + \phi_{amp}^2 + \sigma_0^2} \tag{5}$$

Non-ergodic GMMs include some or all of the systematic effects that are practical to model (second group of terms) as part of the modelled effects (first group of terms). For example, in a partially non-ergodic model with non-ergodic site terms, the average site term, $\delta S2S_s$, becomes part of the median model and the GMM can be written as:

$$\ln(Y_{es}) = [\mu_{erg}(M, R, S, \ldots) + \delta S2S_s] + [\delta L2L_e + \delta P2P_{es}] \tag{6}$$

$$+[\delta B_e^0 + \delta WSP_{es} + \delta AMP_s] + \delta_{0,es}$$

In this case, there is a reduction in the aleatory variability because the average site term is modelled. The total standard deviation of the site term, ϕ_{S2S}^2 in Eq. (5), is no longer treated as aleatory variability. There will be an increase in the epistemic uncertainty in the median because the $\delta S2S_s$ term needs to be estimated from the limited data at the site. For this case, the 'single-station' sigma is given by Eq. (7):

$$\sigma_{SS} = \sqrt{\tau_{L2L}^2 + \phi_{P2P}^2 + \tau_0^2 + \phi_{WP}^2 + \phi_{amp}^2 + \sigma_0^2} \tag{7}$$

In a fully non-ergodic GMM, the systematic source and path effects are also modelled and the GMM is given by Eq. (8). There is a greater reduction in the aleatory variability and standard deviation for a single path is given by Eq. (9).

$$\ln(Y_{es}) = [\mu_{erg}(M, R, S, \ldots) + \delta S2S_s + \delta L2L_e + \delta P2P_{es}] \tag{8}$$

$$+[\delta B_e^0 + \delta WSP_{es} + \delta AMP_s] + \delta_{0,es}$$

$$\sigma_{SP} = \sqrt{\tau_0^2 + \phi_{WP}^2 + \phi_{amp}^2 + \sigma_0^2} \tag{9}$$

There is also an increase in the epistemic uncertainty in the median because the $\delta L2L_e$, $\delta P2P_{es}$, $\delta S2S_s$ terms all need to be estimated from the limited regional data. There is still aleatory variability, because the non-ergodic GMM only accounts for the average source, path, and site terms and not include all the details in the slip-distribution, hypocentre location, rupture velocity, source-time function, and 3-D velocity structure. An example of the relative sizes of the components of the standard deviation is shown in Table 1 [38]. For this dataset, removing the systematic site term leads to a single-station sigma that reduces the ergodic sigma by 10–15%. Removing the systematic path terms and the systematic source terms in addition to removing the site terms reduces the ergodic sigma by 40–50%.

TABLE 1. NOTATION AND EXAMPLE OF THE COMPONENTS OF THE STANDARD DEVIATION OF GROUND MOTION (ADAPTED FROM REF. [38])

Term	Notation for residual	Notation for standard deviation	Standard deviation of peak ground acceleration (ln units)	Standard deviation of pseudo-spectral acceleration at T=1 s (ln units)
Systematic source term for a given location (deviation from the magnitude scaling in the reference GMM)	$\delta L2L_e$	τ_{L2L}	0.254	0.323
Systematic site term (deviation from site scaling in the reference GMM)	$\delta S2S_s$	ϕ_{S2S}	0.259	0.364
Systematic path term (deviation from distance scaling in the reference GMM)	$\delta P2P_{es}$	ϕ_{P2P}	0.401	0.348
Variability of the event terms about the median source term for a location	δB_e^0	τ_0	0.247	0.314
Variability of the path effect about the median path effect	δWP_{es}	ϕ_{WP}	Not estimated	Not estimated
Variability of the site amplification variability about the median site term	δAMP_{es}	ϕ_{Amp}	Not estimated	Not estimated
Within-site variability $\delta WSP_{es} + \delta Amp_s$	δWS_{es}	ϕ_{ss}	0.230	0.294
Inherent randomness	$\delta_{0,es}$	σ_0	-	-
Total (ergodic) sigma	-	σ_T	0.637	0.737
Single-station sigma	-	σ_{SS}	0.583	0.640
Single-path sigma	-	σ_{SP}	0.337	0.430

4.2. EPISTEMIC UNCERTAINTY IN ALEATORY VARIABILITY FOR GROUND MOTION MODELS

In PSHA, epistemic uncertainty is modelled by logic trees. Examples of GMM logic trees for ergodic and non-ergodic GMMs are shown in Fig. 14. As explained in Ref. [37], the logic trees for the ergodic and non-ergodic GMMs are kept separate; they are not to be combined into a single logic tree, because they have significant differences in the level of model simplification leading to a very different aleatory variability and slopes of the hazard curves. Combining the results of these different level of model simplification will not reflect the appropriate epistemic fractiles on the hazard as shown in Section 4.4.

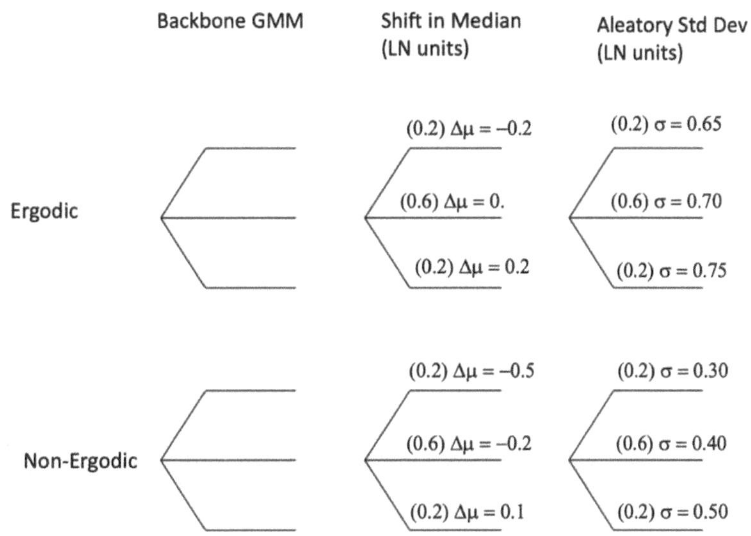

FIG. 14. Comparing the logic trees for median and aleatory variability in ergodic and non-ergodic models (numbers in parentheses represents the weights assigned to each branch).

In this example, the first node of the logic tree represents the epistemic uncertainty in the median for the scaled backbone model approach [39]. This is different to the ground motion logic tree that includes multiple GMMs; instead, the centre body and range of the median predictions of multiple GMMs is covered by a backbone model with different scaling factors. For the ergodic GMM, the epistemic uncertainty in the median is centred on zero and there is a narrow range of ±0.2 due to the large data set used to develop the ergodic GMMs. For the non-ergodic GMM logic tree, we assume that there is some local ground motion data available to constrain the non-ergodic terms, so the central value of the second node is non-zero (because the median includes the site specific terms given in Eq. 8). The epistemic uncertainty of the median for the non-ergodic GMM is larger when compared to the ergodic branch because the site specific source, path, and site terms are more uncertain due to smaller number of data available to constrain the site specific terms.

The second node shows that there is larger aleatory variability for the ergodic GMM as compared to the non-ergodic GMM. In other words, the standard deviations are higher for the ergodic branches. However, the epistemic uncertainty for the ergodic standard deviation is smaller (the standard deviations vary less) than for the non-ergodic GMM because a larger data set is available to constrain the value of the ergodic standard deviation.

A key limitation of using complex models is that the epistemic uncertainty may be very large for sites with limited data for the controlling scenarios. As the amount of data reduces, the range of variations in the second and third nodes of the logic tree given in Fig. 14 will be higher. The question becomes 'is it useful to have such large uncertainties?'

For a normal distribution of the epistemic uncertainty, the mean hazard from the non-ergodic GMM is the same as using an ergodic GMM with the aleatory standard deviation given by the combined aleatory variability and epistemic uncertainty as in Eq. (10).

$$\sigma_{comb} = \sqrt{\sigma_{al}^2 + \sigma_{epi}^2} \tag{10}$$

If there is no data to contain the non-ergodic terms, then the combined sigma will be equal to the ergodic sigma. However, if there is some local data available, then there will be a net reduction in the combined sigma. Therefore, using a non-ergodic GMM does not simply move some of the aleatory variability to epistemic uncertainty. If some local data is available, then there is an increase in accuracy of the GMM.

If there is no site specific data available, the non-ergodic approach can still be used, but epistemic uncertainties in the systematic site, path, and source terms will be large. The combined aleatory and epistemic uncertainties for the ground motion will be the same as for the ergodic GMM. As a result, the mean hazard will be unchanged, but using the non-ergodic approach will show that the epistemic uncertainties in the hazard are much larger than those computed using the ergodic GMM. If the non-ergodic approach is applied to regions without local data to constrain the non-ergodic terms, the hazard also needs to be computed using ergodic GMMs and used to check that the mean hazard from the non-ergodic GMM and ergodic approaches are the same. This serves as a sense check on the implementation of the non-ergodic GMM into the PSHA.

4.3. MODELLING ALEATORY VARIABILITY AND EPISTEMIC UNCERTAINTY IN GROUND MOTION MODELS

Reference [37] describes a framework for tracking the various components of aleatory variability and epistemic uncertainty for both empirical GMMs and numerical simulations. There are three components of the model: aleatory variability of the ground motion, the epistemic uncertainty in the model for the median ground motion, and the epistemic uncertainty in the model for the aleatory variability. The matrix shown in Table 2 (adopted from Ref. [37]) is a useful tool for tracking the terms that contribute to either the aleatory variability or the epistemic uncertainty for GMMs.

The aleatory term in the first row of Table 2 represents the limitations of the simplified models to predict the observed ground motions. It is measured from comparisons of the model predictions to the observed data (or the residuals). The standard deviation of the GMM is an example of the aleatory variability. This aleatory variability is modelled with a PDF which is often truncated at a specified number of standard deviations above or below the median. In general, models with greater simplification will have larger aleatory variability than complex models that include more of the physical behaviours of ground motions. The epistemic uncertainty in the median includes the statistical uncertainty in the estimate of the median due to the limited number of observations. In its simplest form, this is the standard error of the mean. As additional non-ergodic terms are included in the GMM, there are additional terms that have epistemic uncertainty. For example, the epistemic uncertainty in the median site term depends on the number of observations at the site as shown in Fig. 13. There is also the epistemic uncertainty in the reference ergodic GMM that is captured using the traditional logic tree approach with alternative GMMs.

TABLE 2: DEFINITIONS OF UNCERTAINTY TERMS IN GROUND MOTION MODELLING FOR EMPIRICAL GMMS (ADAPTED FROM [37]).

Uncertainty Components	Terms
Aleatory variability	σ ✓ Variability due to effects of physical behaviours not included in the GMM. ✓ Computed from residuals of the GMM.
Epistemic uncertainty in the median	σ_{μ} ✓ Statistical uncertainty in the median due to limited number of observations. ✓ Alternative GMMs captured using a logic tree approach.
Epistemic uncertainty in the size of the aleatory variability	σ_{σ} ✓ Statistical uncertainty in the standard deviation due to limited number of observations. ✓ Alternative models for the aleatory variability captured using a logic tree approach.

In a similar manner, the epistemic uncertainty in the aleatory variability includes statistical uncertainty in the estimate of the standard deviation due to the limited number of observations. The number of observations can be large if a single standard deviation is used for all regions, or it can be small if the standard deviation value is also regionalized, or a site specific value is used. There can also be epistemic uncertainty in the value of aleatory standard deviation as a result of using different datasets when computing the standard deviation [40]. Most models simplify the aleatory variability to be constant for a region, but more complex models allow for the aleatory variability to vary from site to site. For example, sites with stronger topographic or 3-D velocity structure effects may have increased aleatory variability [41].

Typically, a more simplified GMM, such as the global GMMs published in the 1990s, groups data from different regions together to better constrain the median scaling in the GMM, whereas a more complex GMM that distinguishes between different regions, such as the NGA-W2 GMMs, will include additional terms to capture the differences in the median scaling for different regions. Grouping data from different regions leads to larger number of observations to use in the development of the model and more stable estimates of the model parameters (median and standard deviation). This leads to a smaller epistemic uncertainty, but at a cost of larger aleatory variability due to ignoring the effects of systematic differences in the ground motion for the different regions. More complex GMMs will capture some of the physical effects that were ignored in the simplified model, which will lead to less unexplained variability (reduced aleatory variability), but there will be greater epistemic uncertainty in the median GMM because additional terms in the model need to be estimated for the added physical effects. As more data are collected, the epistemic uncertainty in the median GMM will be reduced over time.

4.4. RANGE OF EPISTEMIC UNCERTAINTY

If the ground motion at a site could be measured over tens of thousands of years, then computation of the 'true' hazard at that site is possible. According to the definition of process-based epistemic uncertainty, the range of the final hazard curves (median and full set of fractals) from a PSHA will be broad enough to include the true hazard at the site. If a greatly simplified model (e.g. ergodic GMM) is used, then the epistemic uncertainty range is unlikely to capture the true hazard. The reason for this is that the slope of the hazard curve at long return periods is controlled by the aleatory variability of the GMM. Using an ergodic GMM leads to larger aleatory variability as a result of ignoring site specific and region specific effects that are not consistent with the physical behaviours that apply at the site. As a result, the slope of the hazard curve from an ergodic GMM will be flatter than the true hazard. The epistemic uncertainty range from different ergodic GMMs will be too narrow to capture the true hazard at the site.

As an example, the hazard for ergodic and non-ergodic GMMs for two different sets of non-ergodic terms are shown in Fig. 15. In the first plot, the non-ergodic terms lead to a smaller median ground motion than the ergodic GMM. In this case, the hazard curves are shifted to the left (smaller ground motions) and the curves are steeper. As a result, the ergodic hazard curves are much higher than the non-ergodic hazard curves. In the second plot, the non-ergodic terms lead to a larger median than the ergodic GMM, but with a reduced aleatory variability that results in a steeper slope. In this case, the hazard for the ergodic and non-ergodic models are similar, but they begin to diverge as the return period increases. As a result, the true hazard is not expected to be in the epistemic range of the hazard based on greatly simplified models (e.g. ergodic GMMs).

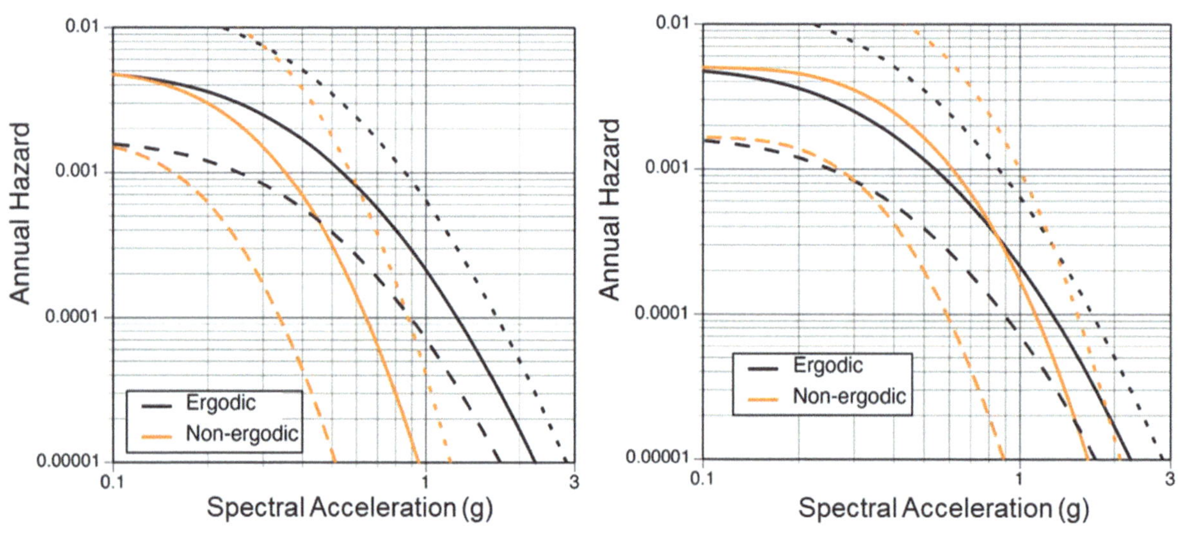

FIG. 15. Median (solid lines) and 5-95% range of uncertainty (dashed lines) in ergodic and non-ergodic hazard curves for a single scenario. The left-hand plot has a negative (below average) non-ergodic term of -0.55 to 0.00 (median = -0.27) and the right-hand plot has a positive (above average) non-ergodic term of 0.00 to 0.55 (median=0.27). The median rate of the scenario is 0.005/yr with an epistemic uncertainty in the rate of a factor of 0.33 and 3.0.

5. METHODS TO DEVELOP NON-ERGODIC GROUND MOTION MODELS WITH NON-ERGODICITY IN THE SITE TERMS

Non-ergodicity in the site terms of GMMs may be estimated by using the ground motion recordings at the site and/or using analytical simulations of the site response. In either case, there are three main steps to be followed: (1) estimating the median non-ergodic site terms, (2) computing the reduction in the aleatory variability, and (3) estimating the epistemic uncertainty in the non-ergodic site terms.

Ergodic GMMs group the sites together based on the site class or based on a continuous metric such as V_{S30}. The average and variability of the site amplification in each group is assumed to apply to all sites in the group; however, each site will have its own average site amplification based on the shear wave velocity structure at the site. Non-ergodic models for the site term account for the difference between the site specific site amplification and the average site amplification for the group. The statistical model of a partially non-ergodic GMM with non-ergodic site terms is given in Eq. (11):

$$\ln(Y_{es}) = \mu_{erg}(M, R, S, \ldots) + \delta S2S_s + \delta B_e + \delta WS_{es} \quad (11)$$

in which δWS_{es} is the site corrected within-event residual and δB_e is the between-event residual. The non-ergodic site term, $\delta S2S_s$, in Eq. (11) is not the full site amplification relative to a rock site. It is the difference between the site specific average amplification and the average amplification in the GMM for the selected reference condition, S.

The reduced aleatory variability, σ_{SS}, is computed by removing the standard deviation of the $\delta S2S_s$ from the total standard deviation of the ergodic GMM (Eq. (7)). Depending on the region, the single-station sigma may be 10–25% smaller than the ergodic sigma as shown in Fig. 16 [26].

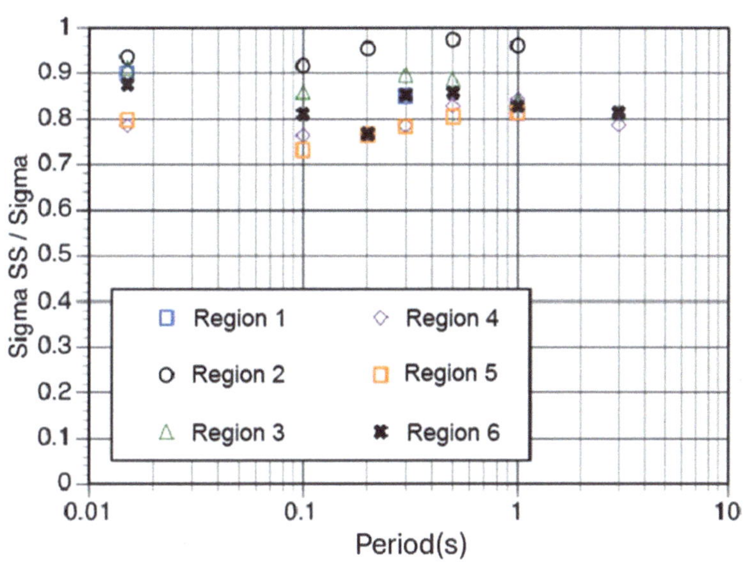

FIG. 16. Comparison of the ratio of single-station within-event standard deviations (Sigma SS) to the ergodic within-event standard deviations (sigma) for different regions. Computed using the ϕ and ϕ_{SS} from Ref. [26] with a constant τ of 0.40.

The σ_{SS} is a regional term, but it can vary from site to site within a broad region. To distinguish between a site specific value and a regional value, the single-station sigma for a particular site is denoted by $\sigma_{SS,s}$ [27]. In this case, the model is non-ergodic in both the median and the aleatory variability.

5.1. PARTIALLY NON-ERGODIC MODELS WITH NON-ERGODIC SITE TERMS BASED ON GROUND MOTION RECORDINGS AT THE SITE

If there are ground motion recordings available at the site, then the residuals relative to a reference GMM can be used to estimate the non-ergodic site term. The simplest approach is to estimate the $\delta S2S_s$ as the mean within-event residuals, δW_{es}, at the site by using Eq. (12):

$$\delta S2S_s = \frac{\sum_{e=1}^{N_s} \delta W_{es}}{N_s} \tag{12}$$

in which N_s is the number of earthquakes recorded at site s. If N_s is small, then the site term needs to be computed by accounting for the small sample size using the maximum likelihood solution given by Eq. (13).

$$\delta S2S_s = \frac{\phi_{S2S}^2 \sum_{e=1}^{N_s} \delta W_{es}}{N_s \phi_{S2S}^2 + \phi_{SS}^2} \tag{13}$$

The terms ϕ_{S2S}^2 and ϕ_{SS}^2 are explained in Table 1. As the N_s becomes large, the $\delta S2S_s$ becomes equal to the mean residual given in Eq. (12). If there are only a few recordings at the site, then the residuals at the site are partitioned into the $\delta S2S_s$ and the δWS_{es} depending on the ratio of ϕ_{S2S}^2 to ϕ_{SS}^2. For a large ϕ_{S2S}^2/ϕ_{SS}^2 ratio, the $\delta S2S_s$ in Eq. (13) approaches to the simple mean residual given in Eq. (12). For a small ϕ_{S2S}^2/ϕ_{SS}^2 ratio and small N, the $\delta S2S_s$ in Eq. (13) approaches zero.

A limitation of this simple approach is that the path effects may be mapped into the site terms if the distance attenuation to the site deviates significantly from the GMM average attenuation. As an example, the residuals of the ground motion from the 2004 M=6.0 Parkfield, California, earthquake relative to an ergodic GMM are shown in Fig. 17. In this case, the event specific attenuation is stronger than the average attenuation in the GMM, so there is a distance trend in the residuals. An NPP site is located at 80 km distance and the mean residual computed for 30–50 km, shown by the black line, is negative for this distance range. The residual at the site reflects both the difference between the site specific site amplification and the GMM site amplification and the difference in the event specific distance attenuation and the GMM distance attenuation. To isolate the site terms, the residuals can be computed relative to an event specific distance scaling. In this case, the site residual relative to the GMM is equal to -1.2 compared to a mean residual of -0.3 relative to the event specific distance scaling (the black line) [42]. This large difference in the residual at the site shows the importance of removing the event specific path effects from the residuals before computing the site terms.

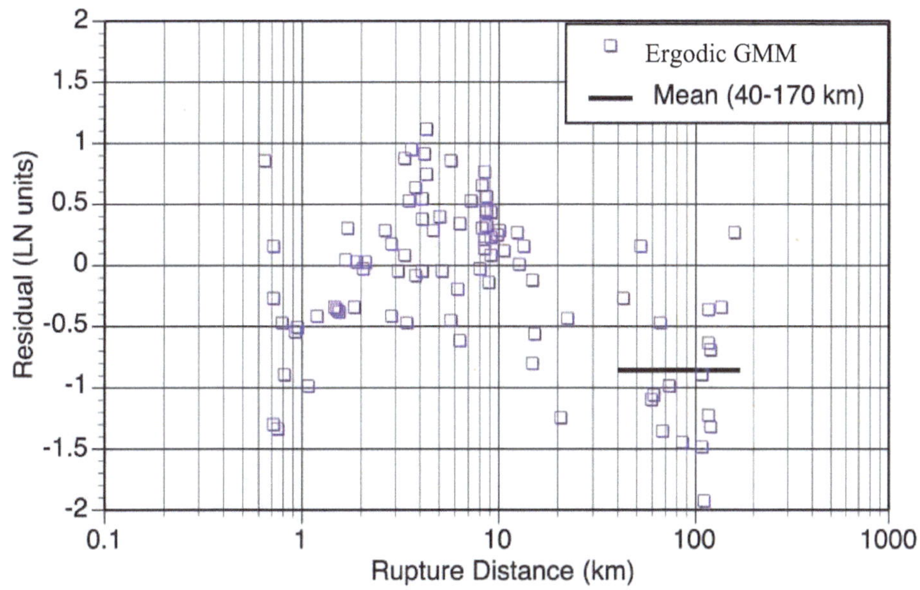

FIG. 17. The within-event residuals of the spectral accelerations at T=0.2 second from the 2004 Parkfield earthquake (adapted from [42]).

Ground motions from small magnitude earthquakes can be used for estimating the linear site term, but there is the issue of the linear amplification of response spectral values depending on the spectral shape. Although the time series of a single degree of freedom (SDOF) system is a linear operation, the response spectral values, given by the peak response of the SDOF, is not a linear operation. The amplification of pseudo-spectral acceleration at a given period depends on the spectral shape, even for strain levels in the linear range. As a result, the amplification from small magnitude earthquakes may not be applicable to large magnitude earthquakes. This is mainly an issue at short spectral periods [43]. The spectral shapes of M3 and M7 earthquakes are compared in Fig. 18(a) and the site factors of the response spectrum are compared to the site factors computed for linear FAS in Fig. 18(b) [44]. At periods less than the period of the peak in the M7 response spectrum (about T=0.2 s), there is a large difference in the response spectral site terms for M3 and M7 events. To avoid this issue, FAS may be used to develop the non-ergodic site terms, but this requires later converting the non-ergodic FAS terms to pseudo-spectral acceleration terms. The most common approach used for the conversion is based on the random vibration theory. An example of this approach is provided in Annex III.

5.1.1. Estimating the reduction in aleatory variability for non-ergodic site terms

To estimate the single-station sigma from empirical data requires datasets with recordings from multiple earthquakes at each site. A common approach is to use a standard mixed-effects regression to estimate the ϕ_{SS} terms. For example, Ref. [26] developed a model for single-station sigma for response spectral values based on data from multiple regions. The ϕ_{SS} values found to be more consistent between different regions then the ergodic ϕ values. This indicates that the main cause for the regional differences in the ergodic ϕ values is the difference in the variability of the site factors.

The standard mixed-effects regression does not account for the spatial correlation of the non-ergodic site terms. Ignoring the spatial correlation leads to a small underestimation of the ϕ_{SS} if the sampling includes data from closely spaced stations. An example of the correction to ϕ_{SS} to account for the effect of spatial correlation is given in Ref. [40].

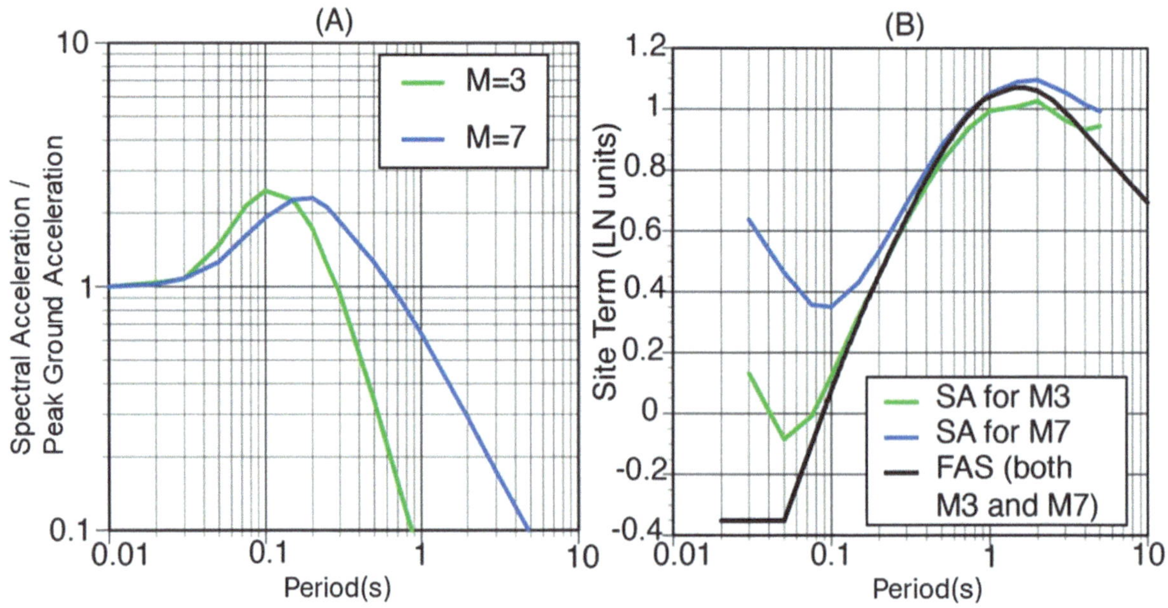

FIG. 18. (a) Comparison of response spectral shapes for M3 and M7. (b) Comparison of the site terms for V_{S30}=300 m/s relative to V_{S30}=760 m/s for M3 and M7 earthquakes. Although the linear FAS site term is same for M3 and M7, the site terms for spectral accelerations for the two magnitudes differ at short periods due to the period range sampled by the spectral acceleration.

5.1.2. Epistemic uncertainty in the non-ergodic site term

The simplest estimate of the epistemic uncertainty in the non-ergodic site term is given by the standard error of the mean of the residuals at site *s* as shown in Eq. (14).

$$SE(\delta S2S_s) = \frac{\phi_{SS,s}}{\sqrt{N_S}} \quad (14)$$

If there are enough recordings at the site, the site specific $\phi_{SS,s}$ can be computed from the standard deviation of the residuals at the site. If there are few recordings at the site, then the $\phi_{SS,s}$ can be approximated by the regional ϕ_{SS} value. An example of the standard error of δS2S is shown in Fig. 13(a). For less than 10 recordings at a site, the epistemic uncertainty grows rapidly as the number of recordings is reduced (steep part of the curve in Fig. 13(a)). However, even with only a few recordings at the site, a more accurate estimate of the site term can be made by including the site specific data.

5.2. PARTIALLY NON-ERGODIC MODELS WITH NON-ERGODIC SITE TERMS BASED ON SITE RESPONSE CALCULATIONS

In many cases, there are no ground motion recordings available at the site. In these cases, the median non-ergodic site term can be estimated using the results of analytical modelling of the site amplification. The most common approach in current practice is to compute the ground motion for a reference rock site condition using the generic single-station sigma in the PSHA and then propagate the rock ground motion through the site specific soil profile. Generic single-station sigma models have been developed from empirical studies (e.g. Refs [27, 41]). An example of generic σ_{SS}, including the epistemic uncertainty, is shown in Fig. 19.

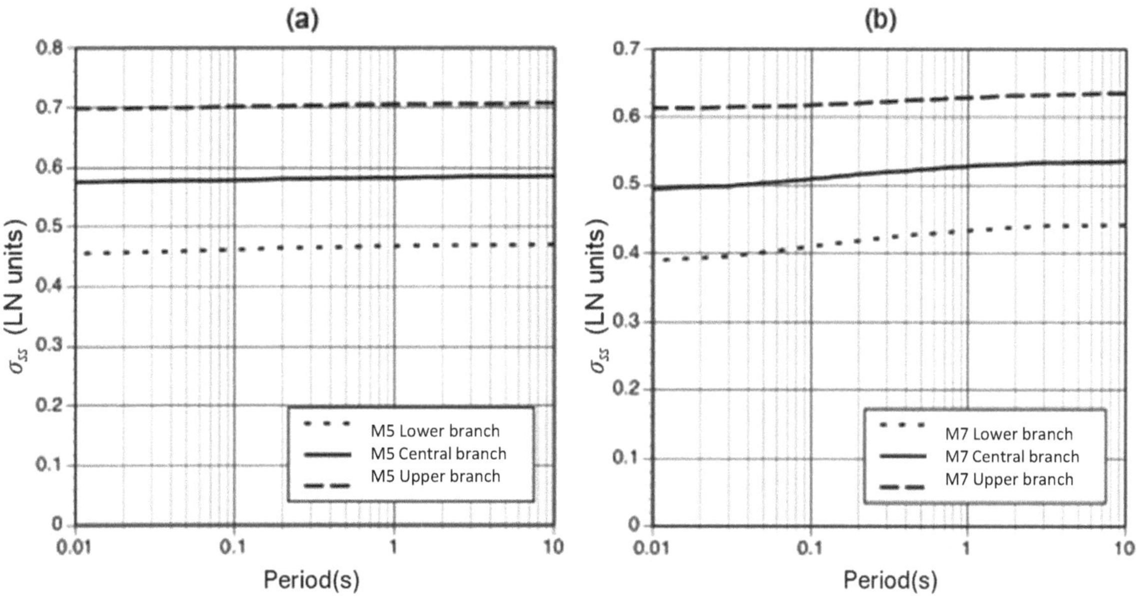

FIG. 19. σ_{ss} model from the Southwestern US study for M5 (a) and M7 (b) (adapted from [40]).

One approach is to define the reference site condition as a hard rock site condition such as Vs=3000 m/s (typical for NPPs in eastern United States of America) or Vs=2000 m/s (used for [45]). In this case, the site response is computed as the amplification for the site specific profile over a hard rock half-space, and the analytically estimated $\delta S2S_s$ term is the traditional site amplification term for soil over rock.

An alternative approach is to define the reference site condition for a lower V_{S30} value. This approach has the advantage that it allows the use of empirical GMMs for a site condition that is better constrained by the empirical data, such as V_{S30}=760 m/s. In this case, the analytically estimated $\delta S2S$ term is the difference between the site amplification for the site specific soil properties (Vs and non-linear material properties) and the site amplification embedded in the GMM for the selected reference site condition. To compute the difference, the site response is computed twice: once for the site specific Vs profile and once for the Vs profile for the site amplification contained in the GMM for the selected reference profile. The same input motion is used for the two profiles and the ratio of the computed surface motions from the two profiles is used to scale the reference rock ground motion from the PSHA. This is similar to the host-to-target correction approach for adjusting GMMs to a region different from the region for which the ground motion data was measured. This approach was used in Ref. [29].

The key issue is that the V_{S30} is not a fundamental parameter for site amplification. Rather, it is an index for the average Vs profile associated with the V_{S30}. The 3-D velocity profile is the physical feature that controls the site amplification, and the full 1-D Vs profile better represents the physical parameters that control site amplification as compared to V_{S30}. The empirically based V_{S30} scaling in most GMMs reflects the correlation between the Vs in the top 30 m and the deeper Vs profile represented in the database used to develop the GMM. This correlation is based on the V_{S30} being measured from the surface and will be different if the V_{S30} is measured at depth from the top of the rock layer. Selecting a V_{S30} value for 'rock' is really selecting the entire Vs profile implicit in the V_{S30} scaling of the GMM. For example, for a reference site condition with V_{S30}=760 m/s in California, there is a shallow gradient, and the average Vs at the surface is about 300 m/s, not 760 m/s (Fig. 20).

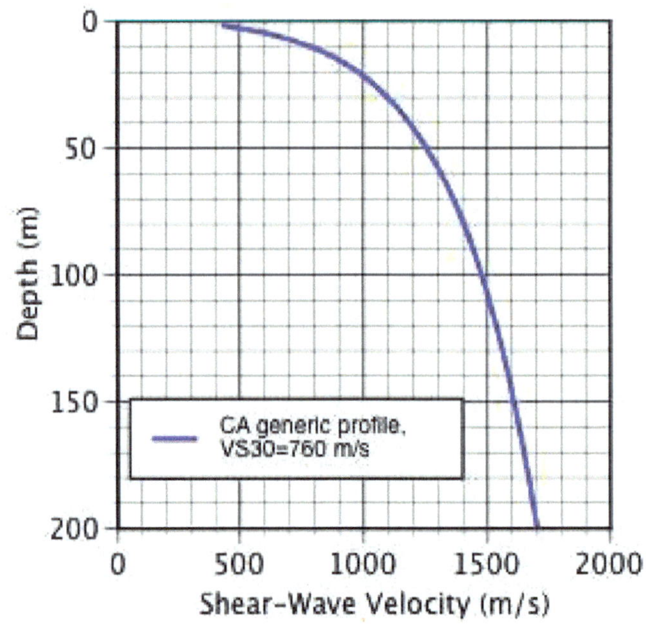

FIG. 20. Example of the shear wave velocity profile for a site with V_{S30}=760 m/s in California. The V_S at the top of the profile is much less than 760 m/s, showing the inconsistency of applying the reference motion at a depth at which V_S=760 m/s.

5.2.1. Epistemic uncertainty in the non-ergodic site term

The epistemic uncertainty in the value of $\delta S2S$ is estimated by computing the site amplification for a range of soil properties that represents the uncertainty in the Vs profile and non-linear soil properties. Similar to traditional site response studies, the site response modelling for non-ergodic site effects can be based on 1-D velocity profiles (e.g. Ref. [26]), 2-D velocity profiles (e.g. Ref. [45]), or 3-D velocity profiles (e.g. Ref. [46]). Either equivalent-linear or non-linear methods can be used depending on the strain levels. The input ground motions can be specified as time series or as FAS combined with random vibration theory.

When performing analytical 1-D, 2-D, or 3-D simulations to support the non-ergodic approach, an appropriate model of the subsurface (velocity) structure is necessary to capture the physics of the subsurface. The subsurface structure itself can be described with three main domains, which decrease in level of complexity with depth: (1) shallow soil layers, (2) deep sedimentary layers below the seismological bedrock (which usually defines the interaction horizon or seismic bedrock above which engineering soil amplification is modelled in more detail), and (3) a crustal structure, which influences the characteristics of the seismic wave propagation.

The main parameters of the upper subsurface structure model to be used for strong ground motion simulation include the density, the shear wave and compression wave velocity profile, and the individual layers with corresponding thickness to represent the stratified structure. An appropriate representation of the seismic velocity structure model includes different models to account for the epistemic uncertainty of the subsurface layers that capture their associated variability in a heterogenous medium as shown in Fig. 21 [45]. Similarly, the uncertainty in the physical material properties is accounted for by considering the range of applicable models.

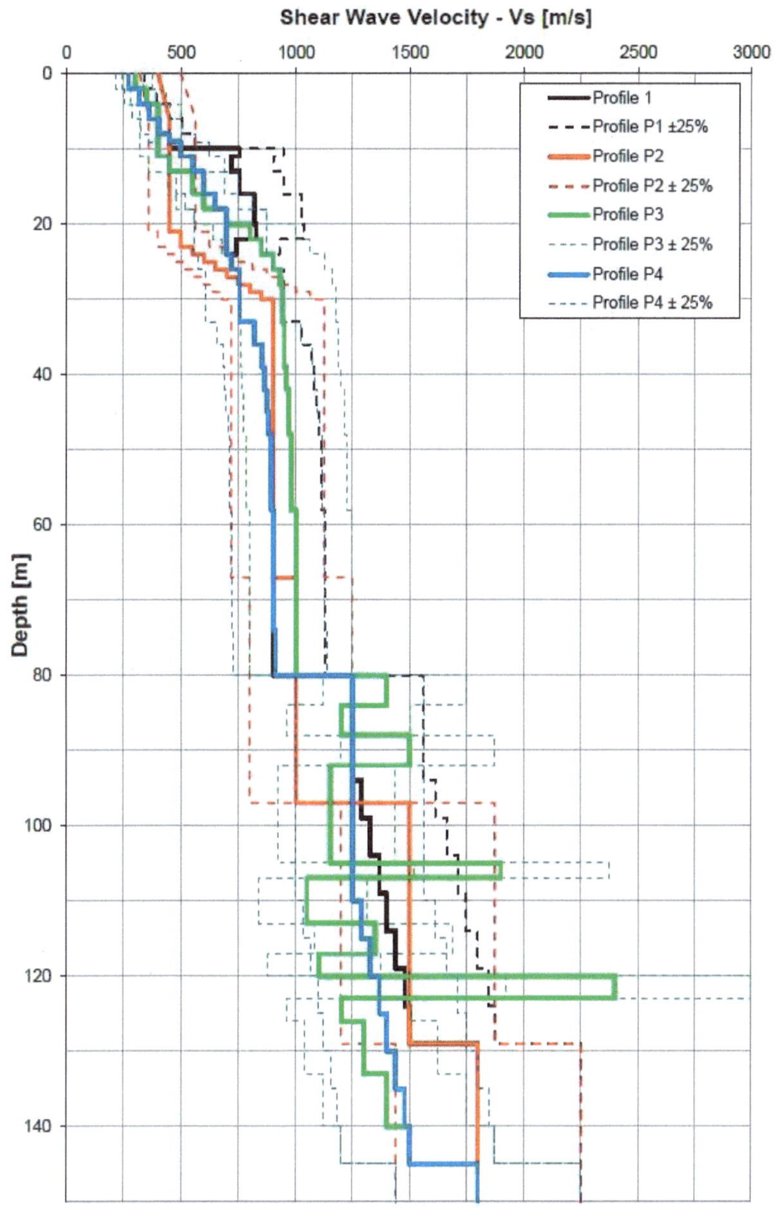

FIG. 21. Example of alternative shear wave velocity (Vs) profiles and associated variability (taken from [45] with permission).

The consideration of alternative models for the velocity profile and non-linear material properties modelling results in a distribution of site amplifications as shown in Fig. 22 [45]. Thus, it is crucial to have sufficient information about the domain to model in the simulations and include the appropriate uncertainty for the specific site. A single model or the velocity profile and non-linear material properties would not capture the possible range of the site parameters. In the case of absence of detailed information to constrain the input parameters, a broader range of alternative input parameters are used for the simulations. The approach to adequately sample the epistemic uncertainty of the site amplification due to the uncertainty input parameters is described in Ref. [47]. A second part of the epistemic uncertainty in the site amplification is model error, which is also described in Ref. [47].

Accurate models for the lower frequency range can be obtained when the deeper soil/bedrock is also modelled in the site response analyses. Otherwise, the model will not capture the long-period features and the epistemic uncertainty in the amplification will be underestimated in this frequency range. For example, the amplification shown in Fig. 22 only captures the range of amplification above 1 Hz. The range of site amplification factors from a limited sample of input profiles and parameters may not cover the uncertainty in site amplification factors in all frequencies [47]. A better approach is to parametrize the epistemic uncertainty in the site amplification factors rather than the input profiles and parameters. The modelling error resulting from 1-D analytical analysis is also included in the epistemic uncertainty of site specific site amplification factors in Ref. [47].

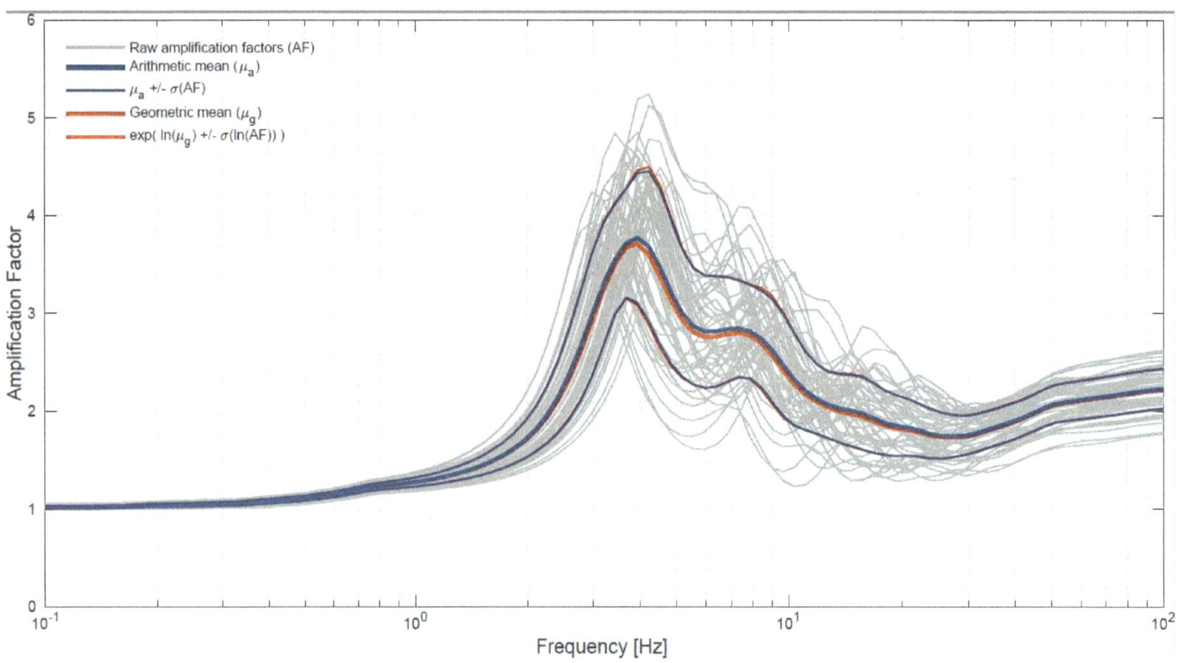

FIG. 22. Example of mean site amplification function and corresponding epistemic fractiles (taken from [45] with permission).

5.2.2. Estimating the reduction in aleatory variability for non-ergodic site terms

An important issue for the analytical modelling is the treatment of the variability of the site response from different input motions. In some applications, this is treated as additional aleatory variability that is combined with the σ_{SS} [48], which may lead to double counting of the aleatory variability [49]. The σ_{SS} models remove the effect of the δS2S from the ergodic standard deviation as shown in Eq. (15).

$$\sigma_{SS} = \sqrt{\sigma^2 - \phi_{S2S}^2} \tag{15}$$

The site amplification at site "s" from earthquake "e", Amp_{es}, can be described by the sum of the median difference in the amplification given by the ground motion model and the site specific amplification, $\delta S2S_s$ (Eq. 16). Variability of the amplification, δAmp_{es}, is related to different input motions and 3-D effects due to different azimuths and incidence angles of the input motions (Eq. 16).

$$Amp_{es} = \delta S2S_s + \delta Amp_{es} \tag{16}$$

The ϕ_{S2S} term is the standard deviation of the $\delta S2S_s$. The ϕ_{amp} is the standard deviation of δAmp_{es} (Table 1). Removing the ϕ_{S2S} from σ does not remove the ϕ_{amp} component, so the σ_{SS} term includes a global estimate of ϕ_{amp} . If the variability of the site response from the site specific site response calculations is included in the hazard calculation, then there will be some double counting of the variability. Three approaches for avoiding the double counting are described in [49]. In the first approach, the site specific ϕ_{amp} is assumed to be equal to the global estimate as shown in Eq. (17).

$$\phi_{amp,s} = \phi_{amp,g} \tag{17}$$

This approach was used in [26]. A limitation of the approach is the ϕ_{amp} may depend on the level of non-linear site effects. The second approach allows for an increase in ϕ_{amp} for high strains to address the potential for an increase in ϕ_{amp} due to non-linear site effects. The difference in ϕ_{amp} at high input ground motion levels from the ϕ_{amp} at low input ground motion levels, representative of the empirical data used to develop σ_{SS} is used to increase ϕ_{amp} as shown in Eq. (18).

$$\phi_{amp_{add}}(PSA) = \begin{cases} \sqrt{\phi^2_{amp,s}(PSA) - \phi^2_{amp,s}(PGA = 0.1g)} & for\ \phi_{amp,s}(PSA) > \phi_{amp,s}(PGA = 0.1g) \\ 0 & for\ \phi_{amp,s}(PSA) \leq \phi_{amp,s}(PGA = 0.1g) \end{cases} \tag{18}$$

In this case, the site specific standard deviation is given by Eq. (19).

$$\sigma(PSA) = \sqrt{\sigma^2_{SS} + \phi^2_{amp-add}(PSA)} \tag{19}$$

The third approach is to remove the global value of ϕ_{amp} and include the full site specific estimate of ϕ_{amp} as shown in Eq. (20).

$$\sigma(PSA) = \sqrt{\sigma^2_{SS} - \phi^2_{amp-g} + \phi^2_{amp-s}(PSA)} \tag{20}$$

The third approach is the most accurate, but it requires an estimate of ϕ_{amp-g} which is not given in most GMMs. Based on borehole data and site response calculations, ϕ_{amp-g} is found to be about 0.25–0.35 ln units by Refs [48], [50] for frequencies above 0.5 Hz.

6. METHODS TO DEVELOP NON-ERGODIC GROUND MOTION MODELS WITH NON-ERGODICITY IN THE PATH TERMS

As discussed in Section 2, the path effects in GMMs are modelled by two terms: the geometrical spreading term (log (R)) and the anelastic attenuation term (linear R). An issue for the path effects is that they depend on both the site and source location. This makes the parameterization of the 'separation distance' between two paths more complex.

Reference [24] used the variable coefficient model (VCM) to include the isotropic path effects for the geometrical spreading terms and anelastic attenuation terms. The geometrical spreading term is often related to saturation of the GMM at short distances and large magnitudes, so including a spatially varying geometrical spreading term can lead to over-saturation in the GMM for some source locations as the GMM is extrapolated to large magnitudes and short distances. For this reason, a spatially varying geometrical spreading term was not included in most of the recent non-ergodic GMMs.

The anelastic attenuation (linear R term) is modelled in non-ergodic GMMs by a cell specific coefficient $\theta_{7,i}$:

$$f_{attn}(\overrightarrow{x_s}, \overrightarrow{x_e}) = \sum_{i=1}^{N_{cell}} \Delta R_i(\overrightarrow{x_s}, \overrightarrow{x_e})\, \theta_{7,i} \quad (21)$$

The f_{attn} term replaces the anelastic attenuation from the ergodic GMM which is modelled by the linear R term, $\theta_7 R$. $\theta_{7,i}$ is the regression coefficient for each cell (i), $\overrightarrow{x_s}$, $\overrightarrow{x_e}$ are the coordinates of the hypocentre and site, and ΔR_i is presented in Fig. 23.

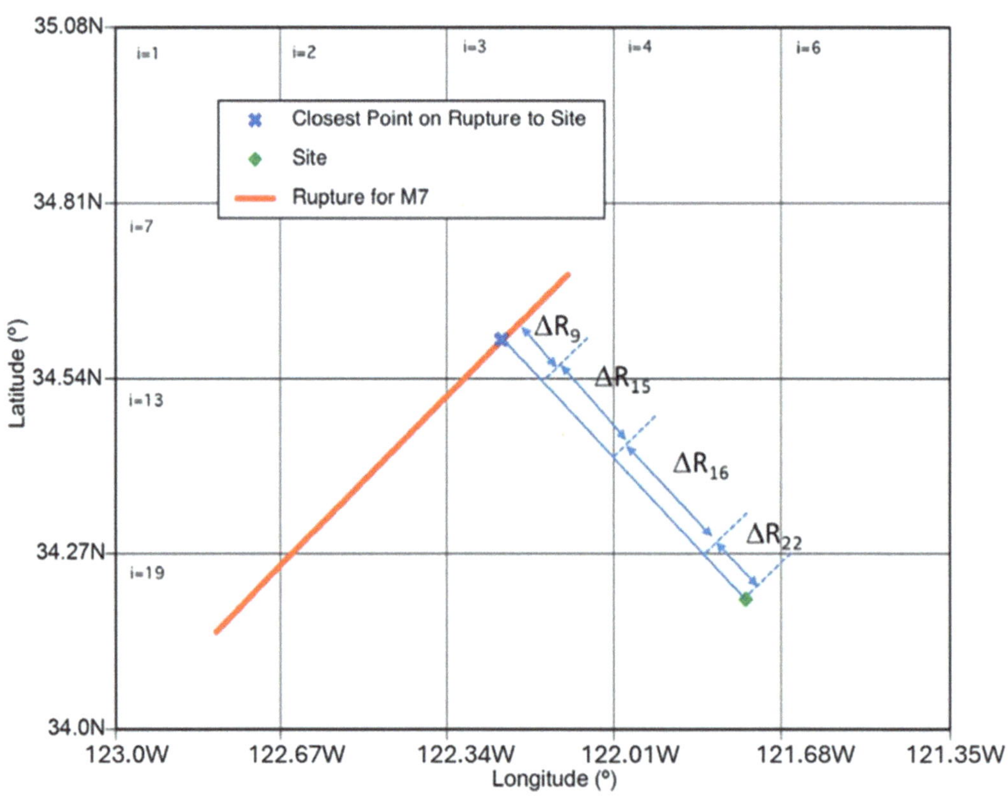

FIG. 23. Schematic illustration of the rupture distance metric (continuous blue line between source and site) and description of the distance computation in each cell for assessing cell specific attenuation (i is the cell number). Horizontal and vertical axes are the coordinates in longitude and latitude in degrees, respectively.

The path term in Eq. (21) can be separated into two path terms that represent two different physical effects: a path term related to the 3-D crustal attenuation structure and a path term related to the 3-D velocity structure. Reference [51] developed a methodology to capture anisotropic path effects due to the 3-D crustal attenuation structure by modelling the linear distance scaling for small individual cells along the ray path rather than using an average value for a region. Reference [52] further extended the cell specific approach of Ref. [51] to estimate systematic path effects based on the posterior distribution of the Bayesian inference, which captures the mean value and epistemic uncertainty of the anelastic attenuation per cell.

In the cell specific approach, the region under study is divided into rectangular cells. For each record in the data set, the length of the ray path within each cell, ΔR_i, is calculated based on a straight line from the site to the closest point on the rupture. The sum of the ΔR_i will be equal to the rupture distance (Fig. 23). For finite ruptures, there is not a single coordinate for the source. One approach for defining the location of a finite rupture is to use the coordinate of the closest point on the rupture to the site, as shown in Fig. 23. Using 3-D simulations of the long-period (T=3 s) response spectra for the Los Angeles region, Ref. [53] showed that this was a reasonable approximation for the path effect, but it has not been tested for short periods. An alternative approach is to compute the weighted average of the path effect over the rupture plane rather than using a single point, with the weights that depend on the distance to the sub-source location. This approach has a better physical basis, but it has not yet been implemented and validated.

The cells approach captures the path specific crustal attenuation effects in the GMM, but significant crustal attenuation effects are mainly limited to short-period ground motion at large distances (e.g. Refs [18], [19], [52]). For short spectral periods (e.g, T=0.2 s), the effect of crustal attenuation begins to be significant at distances greater than 60 km. For long spectral periods (e.g. T=3 s), the distance needs to be greater than 300 km for there to be a significant effect of crustal attenuation. The cell specific approach does not capture the path specific effects on the long-period ground motion at short distances seen in the Cyber-Shake simulations [54]. As a result, the non-ergodic aleatory variability for the recent non-ergodic GMMs (e.g. Ref. [19]) are significantly larger than the empirical estimates for a single-path standard deviation from Ref. [38] and the simulation-based standard deviations estimated by Ref. [36]. This difference indicates that there are additional systematic effects that are not captured by the cell-based approach for path effects in non-ergodic GMMs.

To address this limitation, a second path term that captures path effects related to the 3-D crustal structure is included in the non-ergodic GMMs. Reference [18] proposed a new approach for developing non-ergodic GMMs that captures the effects of the 3-D velocity structure that can be combined with the cells approach for modelling the effects of the 3-D crustal attenuation structure. The new methodology is based on the VCM approach used to develop current non-ergodic GMMs but applied in an iterative manner. The non-ergodic GMM is given in terms of non-ergodic adjustment terms that are added to a reference ergodic GMM for a given source and path pair.

6.1. STATISTICAL APPROACHES FOR FULLY NON-ERGODIC GM MODELS (SITE, PATH, SOURCE)

The functional form of a fully non-ergodic GMM is given in Eq. (22), in which the path term given in Eq. (5) is separated into the 3-D crustal attenuation structure-related term, $\delta P2P_Q$, and the 3-D velocity-structure-related term, $\delta P2P_V$. The inherent randomness due to chaotic effects and quantum effects is assumed to be small and is ignored for this discussion. Also, the site and path aleatory terms are combined into a single within-site within-path aleatory term, $\delta WSP_{es} = \delta Amp_s + \delta WP_{es}$.

$$\text{Ln}\, Y_{es} = GMM_{med}(M, R, S) + \delta L2L(\vec{x}_{e-hy}) + \delta S2S_s(\vec{x}_s) + \delta P2P_Q(\vec{x}_{e-cp}, \vec{x}_s) + \delta P2P_V(\vec{x}_{e-cp}, \vec{x}_s) + \delta B_e + \delta WSP_{es} \tag{22}$$

in which $\vec{x}_{e-hyp}$ is the coordinate of the hypocentre, $\vec{x}_{e-cp}$ is the coordinate of the closest point on the rupture to the site, and $\vec{x}_s$ is the coordinate of the site.

The non-ergodic terms given in Eq. (22) can be estimated from the available ground motion data using a Gaussian process (GP) regression or machine learning methods such as deep learning. VCM based on GP regression is described in detail in Ref. [19], which has been widely used in developing non-ergodic GMM in the last few years. The GP model is conducted in two main operations: estimation of the hyperparameters and forward prediction of the non-ergodic terms.

The first operation is the estimation of two hyperparameters for each non-ergodic term: the variance and the spatial correlation length. In Eq. (22), there are four non-ergodic terms and two aleatory terms. This leads to a total of 10 hyperparameters (4 variances and 4 correlation lengths for the non-ergodic terms, and 2 variances for the aleatory terms). The hyperparameters can be estimated in a single step or they can be estimated in multiple steps to have more stable results to reduce the trade-offs in the non-ergodic terms. In the multi-step approach, the VCM is first used to estimate the hyperparameters for $\delta L2L(\vec{x}_e)$ and $\delta S2S_s(\vec{x}_s)$ and forward prediction is used to compute the non-ergodic source and site terms and the remaining within-site residuals, δWS. Next, the cell approach is used to estimate the coefficients for the cells and to compute $\delta P2P_Q$. Finally, a second VCM is used to estimate the hyperparameters for the $\delta P2P_V$.

The second operation estimates the median non-ergodic terms for a specific source and site location pair $(\vec{x}_e, \vec{x}_s)$ given the hyperparameters and the observed ground motion data. This is called the prediction phase of the GP model.

The approach presented here can be used with either empirical ground motion data or with 3-D simulations. It provides estimates of the median non-ergodic terms, the reduction in the aleatory variability, and the epistemic uncertainty in the non-ergodic terms.

6.1.1. Example application

An example of a non-ergodic GMM developed using the VCM approach with smoothed FAS data (called EAS) from France is presented in this section. This example includes the $\delta L2L(\vec{x}_e)$, $S2S_s(\vec{x}_s)$, and $\delta P2P_Q(\vec{x}_e, \vec{x}_s)$ terms. The $\delta P2P_V(\vec{x}_e, \vec{x}_s)$ is not included in this example. The median values for the non-ergodic source term and site term for 5 Hz EAS are shown in Fig. 24. A comparison of the median distance scaling for the ergodic GMM and the non-ergodic GMM including the source, site, and path ($\delta P2P_Q(\vec{x}_e, \vec{x}_s)$) terms is shown in Fig. 25. For Site A (Fig. 25), the median non-ergodic GMM falls below the ergodic GMM, whereas for Site B (Fig. 25), the median non-ergodic GMM is above the ergodic GMM. Since the non-ergodic GMM did not include the $\delta P2P_V(\vec{x}_e, \vec{x}_s)$ term, the path effects are only seen at the larger distances.

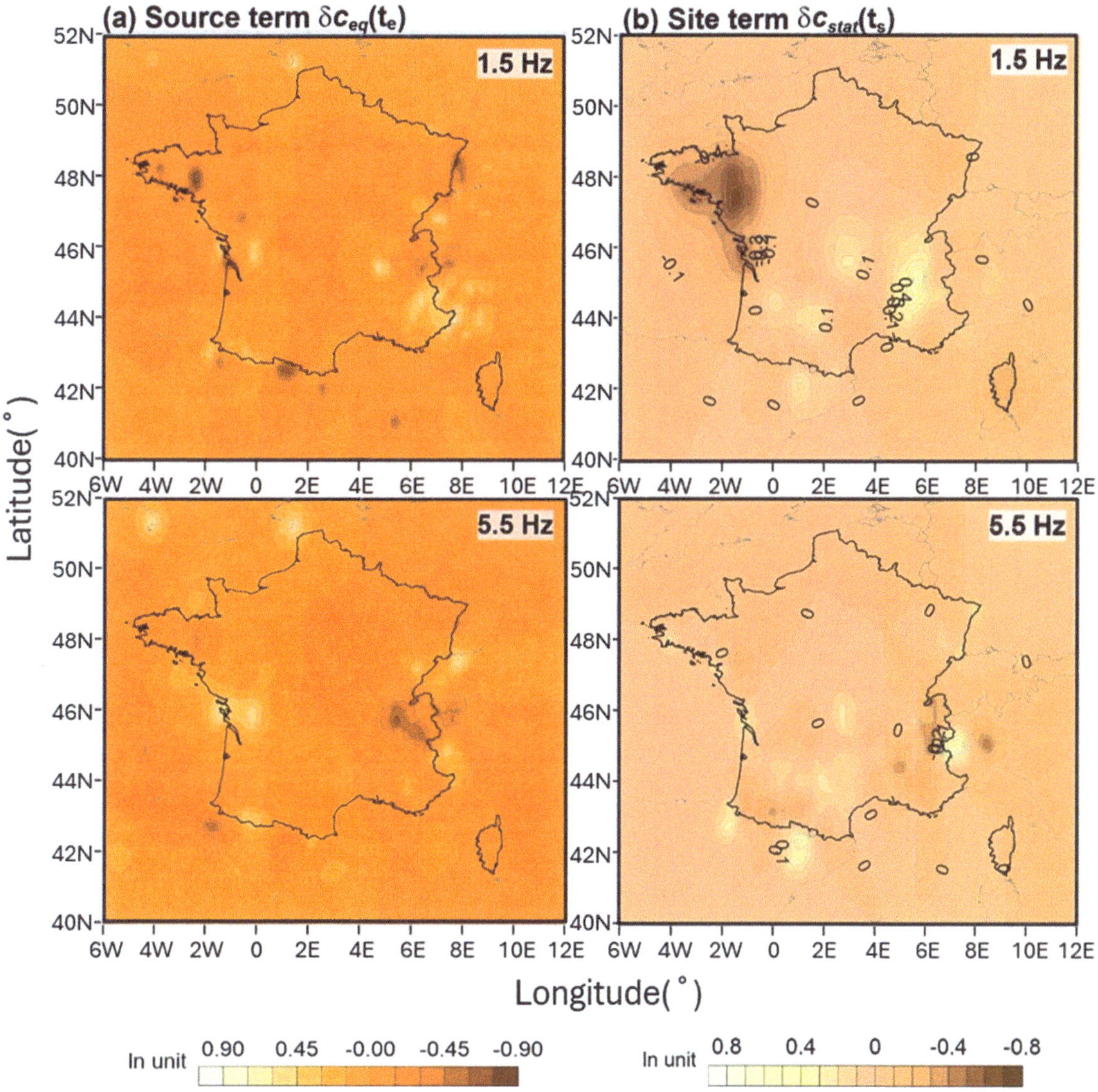

FIG. 24. (a) Source and (b) site terms for a non-ergodic EAS GMM developed for France. Top row shows the source and site terms for 1.5 Hz and the bottom row presents the source and site terms for 5.5 Hz (adapted from [25]). Horizontal and vertical axes are the coordinates in longitude and latitude in degrees, respectively.

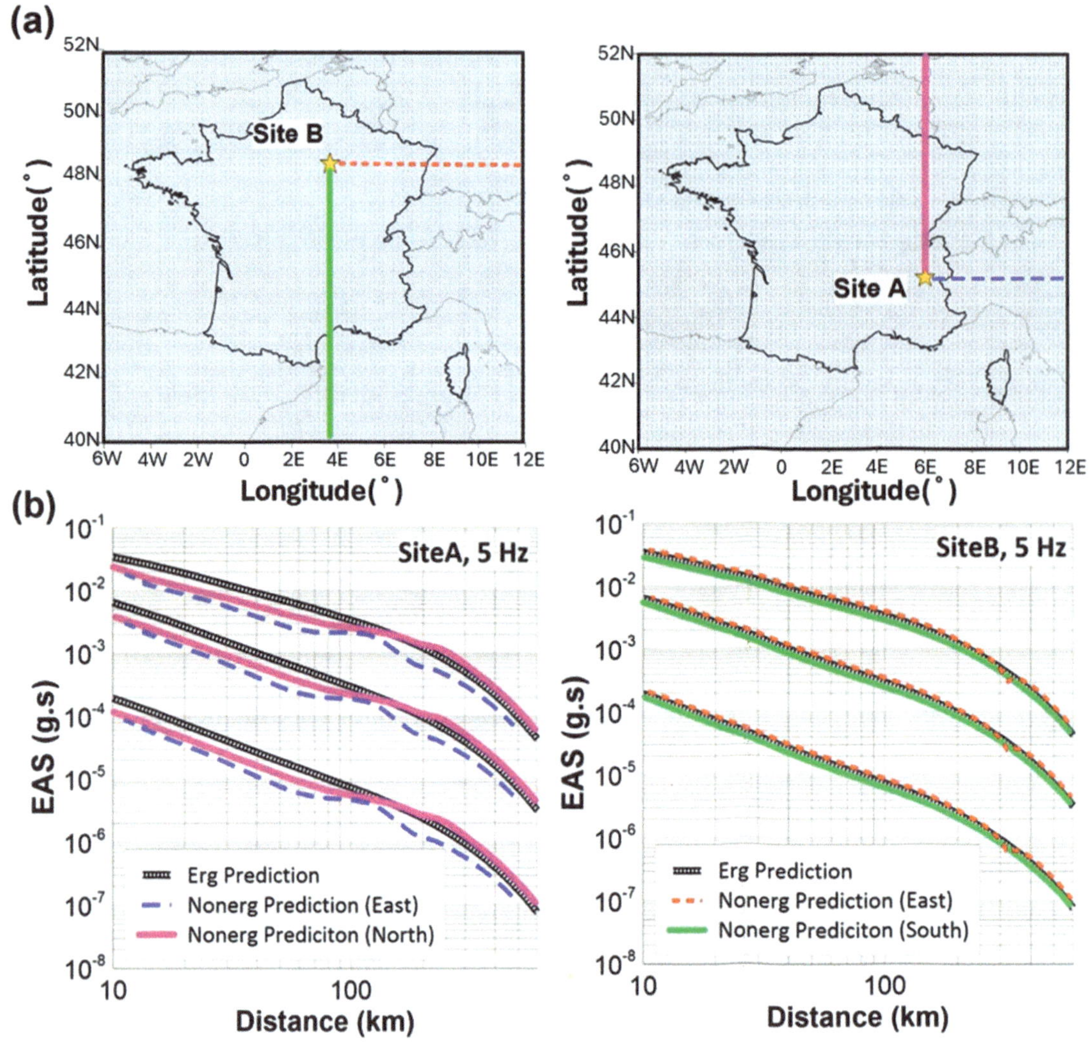

FIG. 25 (a) Location of two example sites, (b) median distance attenuation of the non-ergodic EAS GMM developed for France at 5 Hz at these two example sites (adapted from [25]). Horizontal and vertical axes in (a) are the coordinates in longitude and latitude in degrees, respectively.

The reduction in the aleatory standard deviation is show in Fig. 26. The partially non-ergodic case is for the site terms only. The non-ergodic case leads to a large reduction in the aleatory variability. The epistemic uncertainty in the net non-ergodic term is shown in Fig. 27. The epistemic uncertainty is smallest in the eastern and southern regions of France with the most seismic activity. The epistemic uncertainty is largest in central and northern France, which is an area with little data.

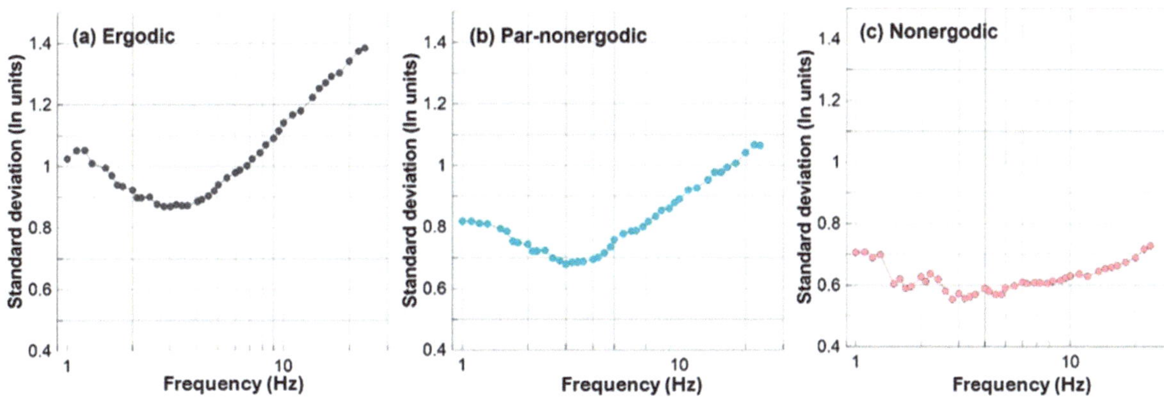

FIG. 26. Reduction of the aleatory standard deviation for (a) ergodic EAS GMM, (b) partially non-ergodic EAS GMM (only includes non-ergodic site term), and (c) non-ergodic EAS GMM for France (adapted from [25]).

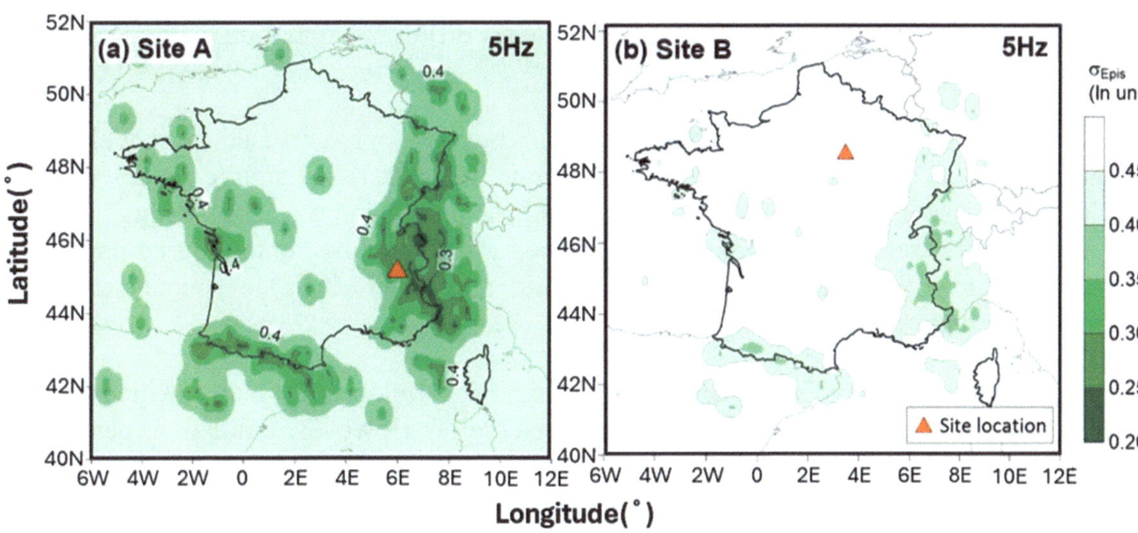

FIG. 27. Epistemic uncertainty for the net non-ergodic term at (a) Site A shown in Figure 25(a) and (b) Site B shown in Figure 25(a) (adapted from [25]). Horizontal and vertical axes are the coordinates in longitude and latitude in degrees, respectively.

7. METHODS TO INCORPORATE GROUND MOTION SIMULATIONS TO NON-ERGODIC MODELS

Physics-based simulation approaches using finite fault models may be used to evaluate the non-ergodic effects on ground motions in the vicinity of the faults, where there are insufficient empirical observations of ground motions. This section discusses how the results of finite fault simulations may be incorporated in the development of non-ergodic GMMs.

A key advantage of the 3-D simulations over empirical ground motion data is that the simulations can generate ground motions with much denser spatial coverage, so the results can be used for fine regionalization of the non-ergodic path and site terms.

7.1.PHYSICS-BASED SIMULATION APPROACHES USING FAULT MODELS

Multiple physics-based ground motion simulation methods have been developed and tested over the last decades. For example, the Southern California Earthquake Centre broadband ground motion platform (SCEC-BBP, [55]) include six different simulation methods proposed by different authors [56]–[62].

SCEC-BBP distinguishes the simulation methods by the approach used in modelling the seismic source and the seismic wave propagation. In the seismic source model used for simulations, the source parameters are determined for three source characteristics: the outer source parameters representing the shape and area of the rupture; inner source parameters describing mainly the heterogeneity of the fault; and other dynamic source parameters describing the rupture processes.

It's important to note that SCEC-BBP is developed for 1-D velocity structures, which is useful for evaluating the ground motion scaling for a broad region. However, simulations based on 1-D velocity structure are not useful for constraining non-ergodic GMMs because the 1-D velocity models reflect average wave-propagation effects and not the non-ergodic path and site effects. For use in non-ergodic GMMs, the simulations need to be based on 3-D velocity models. The source characterization models used in 1-D simulation methods are generally also used for input to 3-D velocity models.

There are three main approaches used for physics-based seismological simulations: semi-empirical, theoretical and hybrid.

The semi-empirical method synthesizes the waveform of a large earthquake from the observed waveforms of small-to-moderate magnitude earthquakes that occurred in the source region of the assumed fault. These observed waveforms are used as elementary waves (Green's functions) that represent the 3-D wave-propagation effects. The advantage of this method is that the empirical Green's functions (EGFs) include the wave propagation effects for a broad frequency band. For evaluations of ground motions for a broad region, the observed waveforms do not need to be at the site of interest, but for non-ergodic GMMs, the observed waveform need to be recorded at or near the site of interest and the hypocentres of the events need to sample the full rupture plane to represent the site specific 3-D effects. This need for ground motion observations at or near the site from earthquakes on the rupture plane of future large magnitude events is a key limitation of the semi-empirical method for use in developing non-ergodic GMMs. While this method can be used if site or source specific EGFs are available, currently, it is not used for non-ergodic GMMs.

The theoretical method computes the site specific ground motion by including the 3-D wave propagation effects using either finite element or finite difference calculations for a 3-D velocity structure. Using a 3-D velocity model, the 3-D effects can be computed for the specific site location and the specific event location. This avoids the need to have observed ground motions at the site of interest from sources on the rupture plane of interest in the semi-empirical method.

The reliable period range of the 3-D simulations depends on the resolution of the geophysical data available to constrain the 3-D velocity structure. Typically, the 3-D velocity structure is constrained for the large scale geological features such as basins. The shallow part of the 3-D velocity model used in the simulations is often modified to limit the Vs to be greater than some minimum value (e.g. 500 m/s) to reduce the calculation time and memory requirements. This limits the reliable period range of the simulation to long spectral periods (e.g. T > 1 s) and can lead to missing effects on the amplitudes of the simulated ground motions for soil sites. Due to this constraint, most of the examples provided in this section are for spectral accelerations at T=3 seconds spectral period. The reliable period range needs to be determined for each simulation based on the information available to develop the 3-D velocity model and on the numerical modelling (e.g. grid size and minimum velocity) used for the simulations. As the computational speed of simulations has improved, theoretical methods have been extended up to 5–10 Hz [63], but the 3-D velocity structure is not well constrained at the shorter wavelengths corresponding to the higher frequencies. That is, the computational capabilities are not the limitation for applying 3-D simulations to higher frequencies; the limitation is the 3-D velocity structure input into the simulations.

The hybrid method combines the long-period ground motion from the theoretical method with the short-period ground motion from the semi-empirical method and stochastics methods (e.g. Ref. [64]). The hybrid method has the advantage of producing broadband seismograms, but the short-period part has the same limitations as the semi-empirical method: to be applicable to non-ergodic GMMs, the EGFs need to be recorded at or near the site of interest from sources on the rupture plane of future events. Therefore, the high frequency ground motions from hybrid method simulations are generally not applicable for non-ergodic GMMs.

It needs to be emphasized that the epistemic uncertainty in the non-ergodic terms is needed for seismic hazard applications. The uncertainty in the 3-D velocity structure is a main part of the epistemic uncertainty of the non-ergodic terms based on the theoretical method; however, this source of epistemic uncertainty is usually not evaluated because it can be a major effort to develop alternative 3-D velocity models and then conduct the 3-D simulations. This is a key limitation of non-ergodic GMMs developed from 3-D simulations.

7.2. CONSTRAINING NON-ERGODIC PATH AND SITE EFFECTS USING SIMULATIONS

The simulations using the theoretical method with a 3-D velocity structure is the main physics-based method used for developing non-ergodic GMMs. These simulations can be used directly in place of a GMM [35], or they can be used to adjust a reference ergodic GMM to include non-ergodic effects. Figure 28 compares the hazard curves for spectral accelerations at T=3 s using an ergodic GMM, simulations based on 1-D crustal structure (using SCEC-BBP), and simulations based on 3-D crustal structure (CyberShake, [65]). In this example, the hazard results using the 3-D simulations is similar to the hazard from the ergodic GMM at short return periods (up to 1000 years) but is lower than the ergodic hazard at long return periods due to the reduced aleatory variability. This example also shows that the simulations based on the 1-D crustal model do not capture the non-ergodic path terms.

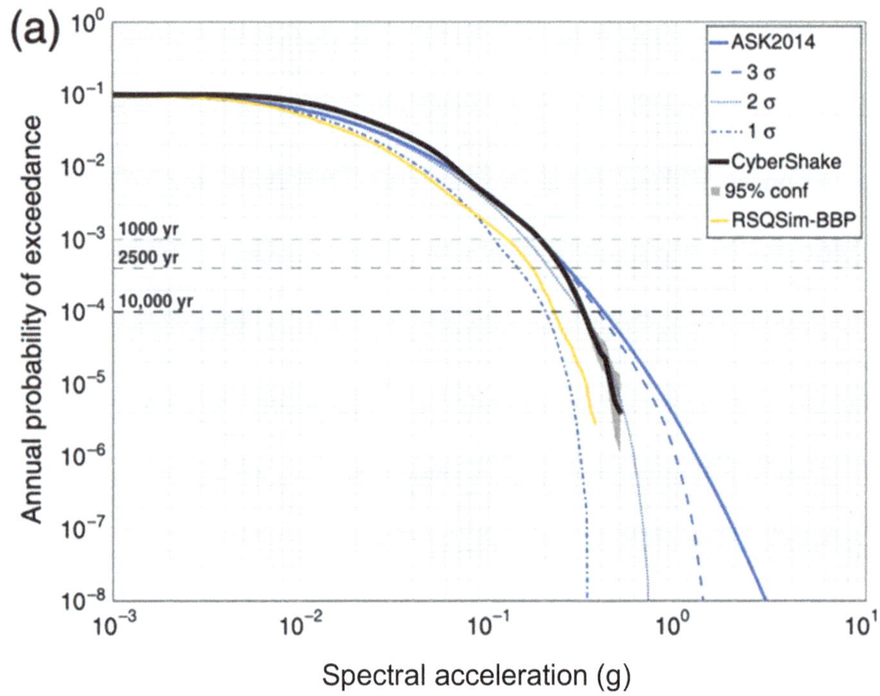

FIG. 28. Comparison of hazard curves for spectral accelerations at T=3 s spectral period for an example site in Southern California using an ergodic GMM, simulations based on a 1-D crustal model, and simulations based on a 3-D crustal model (taken from [35] with permission).

To adjust a reference ergodic GMM to account for non-ergodic path effects, the relative differences in the ground motions due to 3-D path effects can be used rather than using the absolute amplitude of the simulations [66]. The overall amplitude of the ground motions from the ergodic GMM are maintained by adding path specific adjustment terms and a reduction in the aleatory variability. The main steps of this procedure are summarized as follows.

1. If needed, adjust the basin scaling in the reference ergodic GMM to be consistent with the average basin scaling in the 3-D simulations.

2. Compute the residuals of the 3-D simulations (SIM_{es}) relative to the ergodic GMM from Step 1 (GMM_{es}) with a constant term, $C_{SIM}(T)$, as shown in Eq. (23).

$$\delta_{es} = SIM_{es} - GMM_{es} - C_{SIM}(T) \tag{23}$$

   Here, the C_{sim} (T) term is the average difference (in ln units) between the 3-D simulations and the reference ergodic GMM. It reflects the difference in the average source model used in simulations and the source scaling in the GMM.

3. Use the spatial variability of the residuals to estimate the non-ergodic path and site terms. One approach is a Gaussian process regression which uses the spatial correlation and variance of the non-ergodic terms to interpolate the simulation results to any site in the region (e.g. Ref. [66]). The total residual is separated into a non-ergodic path term, $\delta P2P(\vec{x_s}, \vec{x_e})$, a non-ergodic site term, $\delta S2S(\vec{x_s})$, and the remaining within-site within-path residual, δWSP_{es} as shown in Eq. (24).

$$\delta_{es} = \delta P2P(\overrightarrow{x_s}, \overrightarrow{x_e}) + \delta S2S(\overrightarrow{x_s}) + \delta WSP_{es} \quad (24)$$

The $\delta P2P(\overrightarrow{x_s}, \overrightarrow{x_e})$ and $\delta S2S(\overrightarrow{x_s})$ terms are the median non-ergodic adjustment terms that represent systematic path and site effects.

4. Compute the standard deviation of the δWSP_{es} to estimate the non-ergodic ϕ_{SP}. An average ϕ_{SP} over the region can be used or a site specific ϕ_{SP} can be calculated. Using an average ϕ_{SP} over the region is making an ergodic assumption about the ground motion variability. Using a site specific ϕ_{SP} accounts for the differences in the variability of the path and site effects at different site locations.

5. Estimate the epistemic uncertainty in the non-ergodic path and site terms. The epistemic uncertainty in the $\delta P2P(\overrightarrow{x_s}, \overrightarrow{x_e})$ and $\delta S2S(\overrightarrow{x_s})$ terms depend on number of realizations of the source (how stable are the results given a limited number of realizations) and the uncertainty of the 3-D velocity model (how well do we know the 3-D velocity model). For a large number of realizations of the source (e.g. more than 20), the median from the simulations is usually well constrained, and the epistemic uncertainty in the non-ergodic terms is mainly due to the uncertainty in the 3-D velocity structure.

6. With the non-ergodic site and path terms from the simulations, the median non-ergodic GMM is given by Eq. (25).

$$\ln\left(PSA_{NE}(M,R,S,T)\right) = GMM_{erg}(M,R,S,T) + \delta P2P(\overrightarrow{x_s}, \overrightarrow{x_e}) + \delta S2S(\overrightarrow{x_s}) \quad (25)$$

If the simulations are used for their absolute amplitude rather than only for the non-ergodic site and path terms, then the C_{SIM} term is added to the median non-ergodic GMM as in Eq. (26).

$$\ln\left(PSA_{NE}(M,R,S,T)\right) = GMM_{erg}(M,R,S,T) + \delta P2P(\overrightarrow{x_s}, \overrightarrow{x_e}) + \delta S2S(\overrightarrow{x_s}) + C_{SIM} \quad (26)$$

It is important to include the epistemic uncertainty in the estimates of the non-ergodic terms in the application of the non-ergodic GMM. This is done using a logic tree to sample the epistemic uncertainty of the non-ergodic terms as presented in Section 8.

In some applications, the 3-D simulations are computed for a single dominant source rather than for all possible sources to limit the number of simulations. In this case, the path effects cannot be separated from the site effects, and the non-ergodic adjustment is the combined site and path term [66]. In other applications, multiple sources are used in the simulations that sample different path paths. In this case, the site and path effects can be separated [19]. The site term represents the average path effect over all source locations and the path term represents the difference for a specific path from the averaged site term.

7.2.1. Single scenario example from Cascadia

M9 earthquakes on the Cascadia subduction zone were modelled using 3-D simulations for the path and site effects [30]. In all, 30 realizations of the slip distribution and hypocentre for M9 interface events on the Cascadia subduction zone were modelled. In these simulations, only one scenario is considered in the non-ergodic model proposed by Ref. [66]. As a result, the non-ergodic path and site terms were not separated, and the non-ergodic term is the combined path and site term ($\delta P2P(\overrightarrow{x_s}, \overrightarrow{x_e}) + \delta S2S(\overrightarrow{x_s})$).

There are large systematic differences in the site and path effects due to the 3-D velocity structure in the Cascadia region [30]. The spatial distribution of the basin term for the ergodic approach based only on the depth to V_S=2.5 km/s ($Z_{2.5}$) is shown in Fig. 29 [66]. For the ergodic model, the basin terms simply follow the $Z_{2.5}$ contours as shown in Fig. 29(a). The non-ergodic adjustment for the combined path and site effects ($\delta P2P(\overrightarrow{x_s}, \overrightarrow{x_e}) + \delta S2S(\overrightarrow{x_s})$) is shown in Fig. 29(b). When compared to Fig. 29(a), the largest changes are observed at the edges of the basins, which can be either positive or negative depending on the direction of the waves entering the basin.

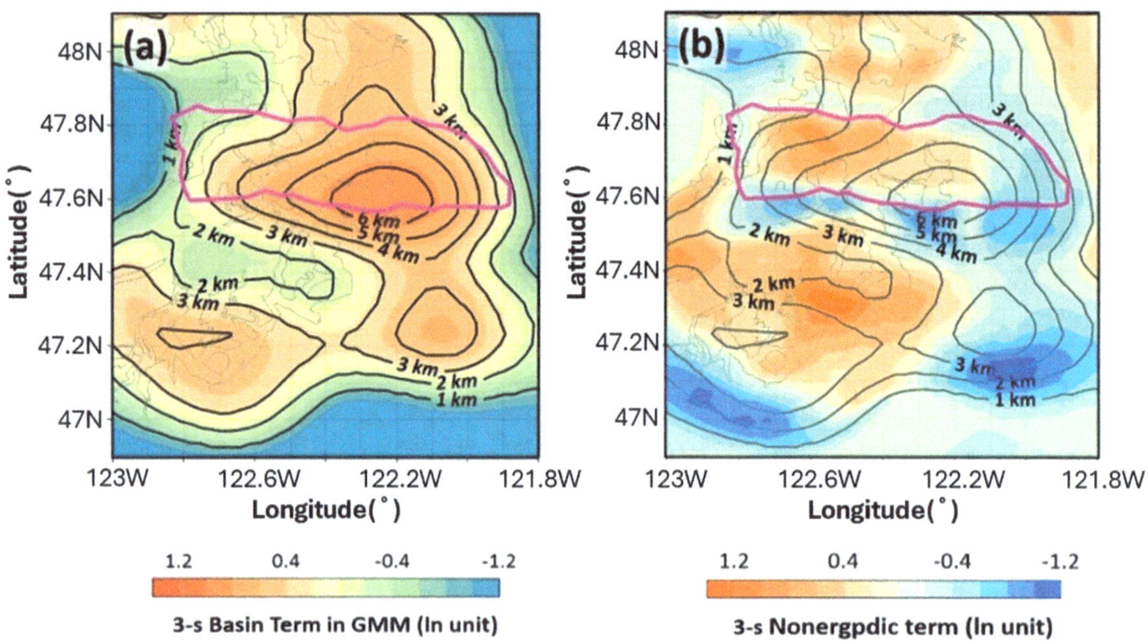

FIG. 29. Comparison of median basin terms for ergodic and non-ergodic GMMs adjusted using 3-D simulations for spectral acceleration at T=3 seconds, the pink line indicates the range of the Seattle basin. (a) Median basin terms based only on $Z_{2.5}$ (ergodic GMM). (b) Non-ergodic adjustments for the combined path and site term (adapted from [66]). Horizontal and vertical axes are the coordinates in longitude and latitude in degrees, respectively.

The reduction in the aleatory variability by including non-ergodic path and site effects is shown in Fig. 30. At the spectral periods of 1.5–4 seconds, the aleatory variability is reduced from about 0.70 to 0.57 in ln units (a reduction of 20%) when compared to the ergodic GMM. The change in the reduction in the aleatory variability for periods greater than 4 seconds is due to the basin 3-D velocity structure which has less of an effect at these longer periods. Most GMMs assume that the aleatory variability is constant for the entire region (i.e. it applies to all sites), but the aleatory variability in a non-ergodic GMM can vary spatially depending on the site specific path and site effects. For example, the ground motion at sites on edges of basins may be more sensitive to the direction of the waves entering the basin, which would increase the variability compared to sites located near the centre of the basin. The site specific variability of the spectral acceleration at T=3 s from the M9 simulation is shown in Fig. 31. The average standard deviation value for the non-ergodic model is 0.56, but at individual sites, the value ranges from 0.49 to 0.7, indicating that for some locations, the aleatory variability for the non-ergodic GMM is as large as the ergodic GMM because the 3-D path effects are more variable at these sites.

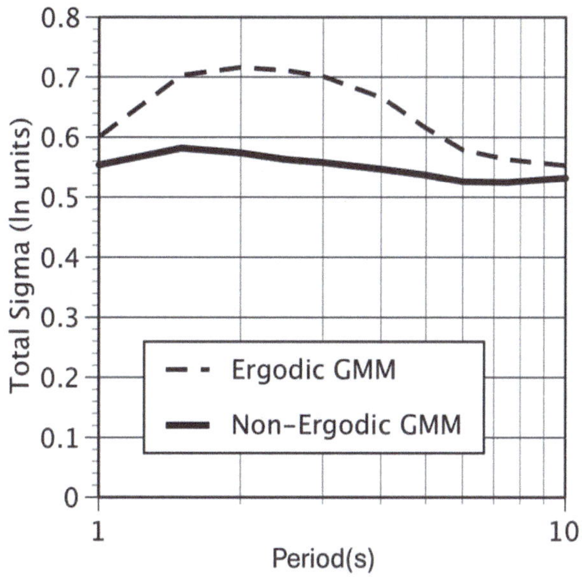

FIG. 30. Total standard deviations from 3-D simulations for M9 events on Cascadia using a global between-event standard deviation value (of 0.47) at all periods (adapted from [66]).

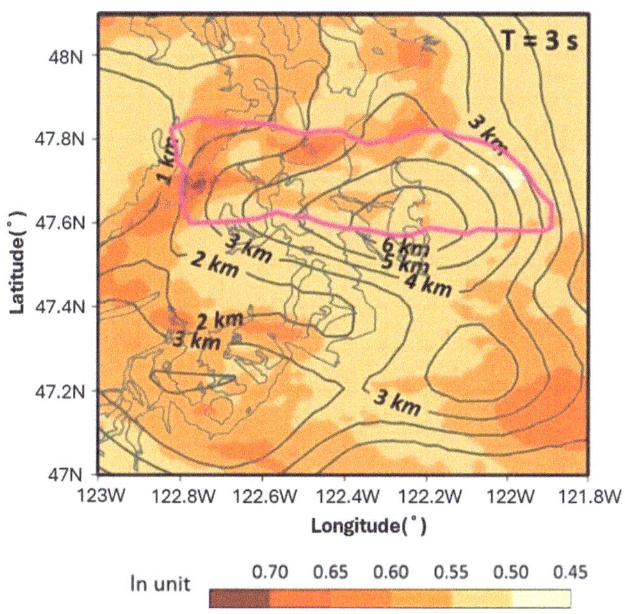

FIG. 31. Example of the spatial variation of non-ergodic standard deviation for spectral acceleration at T=3 s from the M9 simulations for Cascadia. The pink line indicates the range of the Seattle basin (adapted from [66]). Horizontal and vertical axes are the coordinates in longitude and latitude in degrees, respectively.

7.2.2. Single region example from Japan

M8–M9 earthquakes at the Nankai trough subduction zone were modelled with 3-D simulations, considering 400 rupture scenarios with different source areas and slip distributions in Ref. [67]. These simulations are for a source region, but unlike the Cascadia simulations, they include a range of earthquake magnitudes. The probabilistic seismic hazard analysis using the simulation results for Nankai trough subduction zone showed that the amplitude level of ground motion is more significantly affected by the underground velocity structure than the source–site distance.

7.2.3. Multiple scenario example from Southern California

To be able to separate the non-ergodic site terms from the non-ergodic path terms in simulated ground motions, the scenarios modelled in the simulations need to cover a range of wave paths from different azimuths, not just the ray path from a single scenario. This requires a much larger set of scenarios to be simulated. An example of a large set of 3-D simulations for different sources is the CyberShake simulations for the Los Angeles region [54]. CyberShake 3-D simulations for the Los Angeles region include over 100 million simulated ground motions from sources at multiple location with multiple realization of the slip distribution for each source.

As an example, the non-ergodic path terms for spectral acceleration at T=3 s developed in Ref. [18] using a subset of 160,000 simulations are shown for two sites in Fig. 32. The median path term from the non-ergodic GMM depends on the location of the earthquake. Therefore, the distance scaling will vary based on the azimuth to the source (paths A, B, and C, Fig. 32, top row). The distance scaling for the spectral acceleration at T=3 s is shown for the ergodic GMM, the non-ergodic GMM with only the site term (independent of azimuth), and the non-ergodic GMM with both site and path terms in Fig. 32 (bottom row). Figure 32 shows that the effects of the 3-D velocity structure on the wave propagation in different directions can lead to increases or decreases in the ground motion at a given distance. Compared to the ergodic GMM, including the path terms in the non-ergodic GMM can increase or decrease the median ground motion, depending on the azimuth.

The standard deviation of the spectral acceleration at T=3 s from CyberShake simulations using an ergodic approach is 0.56 in ln units. Accounting for only the non-ergodic site terms reduces the standard deviation to 0.516 in ln units, whereas accounting for the non-ergodic path effects and site effects reduces the standard deviation to 0.392 in ln units. Site specific estimates of the standard deviation for the non-ergodic GMM, shown in Fig. 33, can be used in place of the average values. For sites that are more sensitive to the direction of the wave propagation, the non-ergodic standard deviation will be larger than average. In this case, the site specific standard deviations range from 0.34 to 0.50 in ln units.

The largest epistemic uncertainty in the 3-D simulations is due to the uncertainty in the 3-D velocity model. Ideally, we would have simulations for multiple 3-D velocity models to estimate the uncertainty in the non-ergodic terms due to uncertainty in the 3-D velocity model, but there are high computational costs involved with 3-D simulations, and it can be challenging to develop alternative 3-D velocity models that span the epistemic uncertainty range. Without suites of 3-D simulations to sample the uncertainty in the 3-D velocity structure, the epistemic uncertainty of the path terms needs to be estimated. Without suites of 3-D simulations for a range of 3-D velocity models, Ref. [66] assumed an epistemic uncertainty in the path effects of $0.5\sqrt{\phi_{S2S}^2 + \phi_{P2P}^2}$. This corresponds to the 3-D velocity model explaining 75% of the variance of the path effects.

Examples of hazard curves computed using an ergodic GMM, a partially non-ergodic GMM with only non-ergodic site terms, and a non-ergodic GMM with both non-ergodic site and path terms are compared to the hazard computed directly from the CyberShake simulations in Fig. 34. The epistemic uncertainty of the path effects is shown by the thin pink curves that are from 100 branches of the logic tree with each branch having a different map of the path terms. Using the non-ergodic adjustment terms with the reference empirical GMM leads to hazard curves that are similar to the CyberShake hazard results.

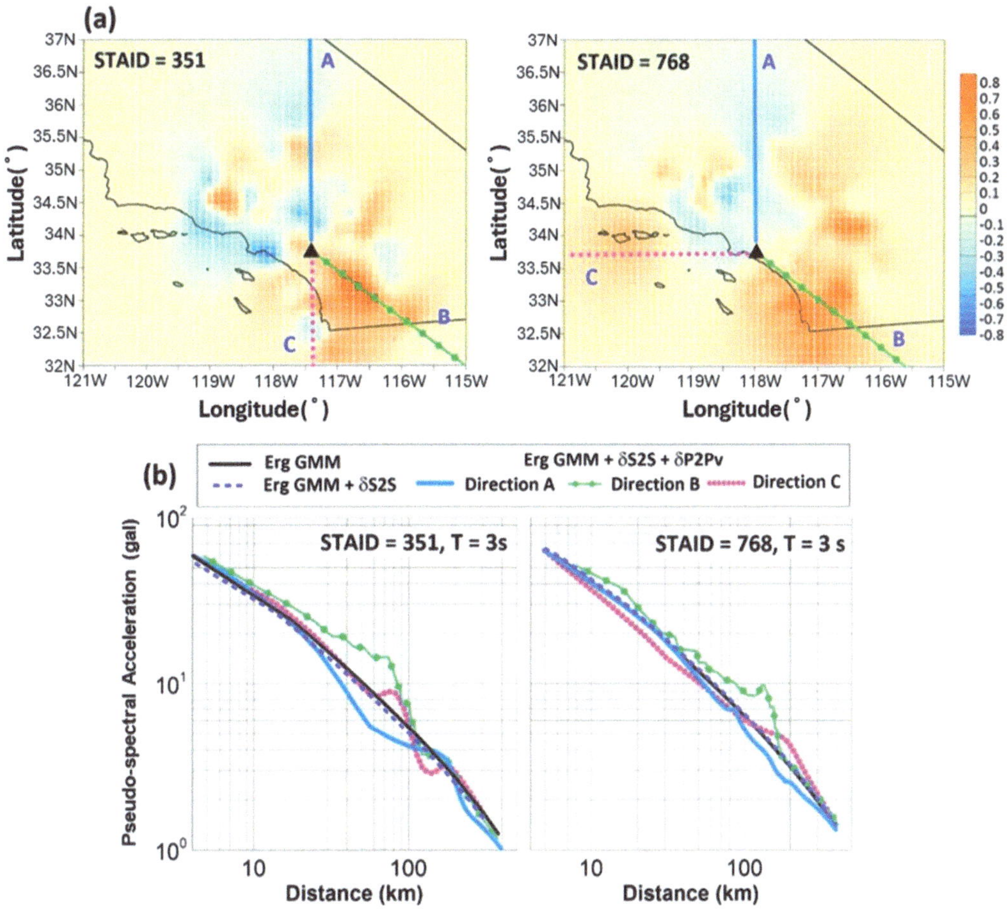

FIG. 32. Example of the azimuthal dependence of the non-ergodic path terms for T=3 s. (a) Maps of the path terms for two station locations, scale bar showing the difference in ground motion in natural log units. Horizontal and vertical axes are the coordinates in longitude and latitude in degrees, respectively. (b) The distance scaling of the PSA in three directions for the ergodic GMM, the non-ergodic GM with the non-ergodic site term, and the non-ergodic GMM with both the non-ergodic site term and path term (adapted from [18]).

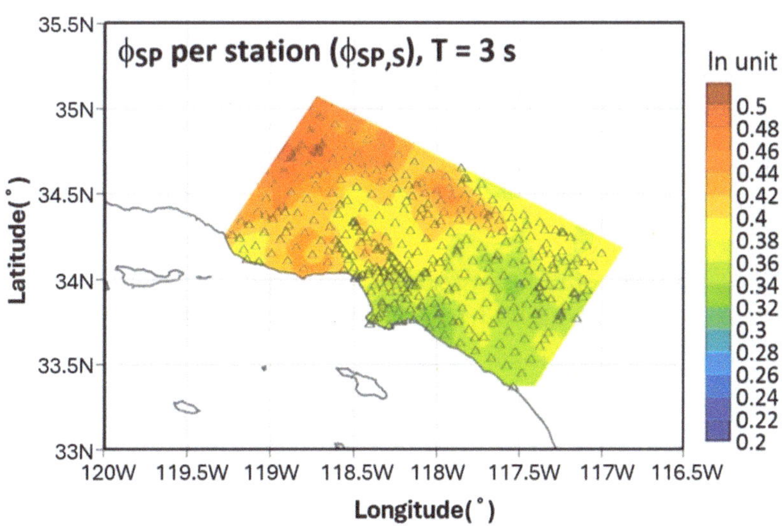

FIG. 33. Example of site specific estimates of the single-path within-site standard deviation. Scale bar shows the value of standard deviation in natural log units. Triangles show the recording stations (adapted from [18]). Horizontal and vertical axes are the coordinates in longitude and latitude in degrees, respectively.

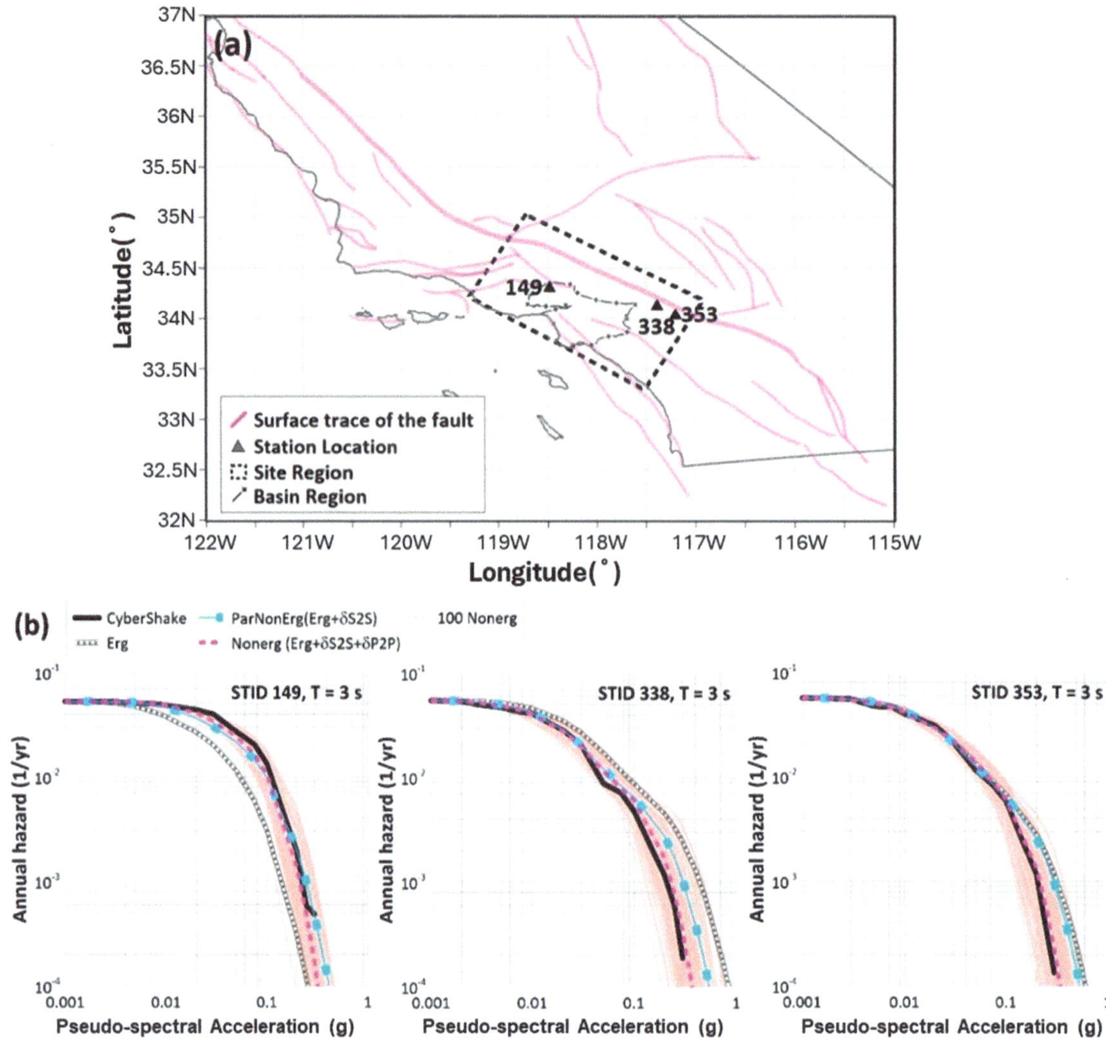

FIG. 34. (a) The locations of example sites, (b) Hazard curves for T=3 seconds based on the 3-D simulations (black curves), the ergodic GMM (dashed grey curves), the partially non-ergodic GMM with only site effects (blue curves), and the non-ergodic GMM with both path and site effects (pink curves). The thin pink lines show the epistemic uncertainty range in the non-ergodic terms (adapted from [18]). Horizontal and vertical axes in (a) are the coordinates in longitude and latitude in degrees, respectively.

7.3.USE OF GROUND MOTION SIMULATIONS TO CONSTRAIN THE NON-ERGODIC SOURCE TERMS

The source models used in simulations are regional average models rather than location specific models. Therefore, they do not address non-ergodic effects for the source other than large scale regional differences, like Japan compared to California. Currently, empirical data is used for non-ergodic source terms rather than simulations.

8. IMPLEMENTATION OF NON-ERGODIC GROUND MOTION MODELS

In ergodic PSHA applications, the way that the median ground motions and the standard deviations are calculated for a site does not depend on the location of the site and seismic source. In other words, the set of site specific parameters utilized in the ergodic GMM (e.g. source-to-site distances) are calculated based on the location of the site, but the ergodic GMM is implemented to estimate the median ground motion and σ_{erg}, independent of the location of the site and seismic source. In non-ergodic PSHA applications, the median ground motion and its standard deviation depend on the coordinates of the site and source, in addition to the set of site specific parameters used in the GMM (e.g. source-to-site distances and site condition). Therefore, to implement the non-ergodic approach in the PSHA calculations, specific adjustments for the source–site pairs, as well as the standard deviation of the epistemic uncertainty in the adjustment need to be included. The following three changes with respect to ergodic PSHA are necessary:

- Median ground motion is modified using the net non-ergodic adjustment terms (sum of the non-ergodic terms for site, path, and source) for each source-site pair;
- The aleatory variability is reduced to account for the inclusion of the non-ergodic terms;
- The epistemic uncertainty of the non-ergodic terms is included in the GMM logic tree with branches that sample the distribution of the epistemic uncertainty of the non-ergodic terms.

These modifications result in a change in the computation of the hazard equation and implementation of the logic tree branches in the standard hazard assessment software as explained in this section.

8.1. HAZARD COMPUTATION EQUATIONS

The application of the non-ergodic approach including epistemic uncertainties in PSHA at a given site implies that many alternative non-ergodic terms will be needed to capture the epistemic uncertainties from sources at different locations. That is, in addition to the standard epistemic uncertainties in the scaling of the median ground motion in ergodic GMMs, the epistemic uncertainty in the path effects for each source that is hazard significant needs to be included.

Different approaches can be used to propagate the epistemic uncertainties of the non-ergodic terms. In a classical ergodic hazard computation, the conditional probability of exceeding a given ground motion level Z is given by Eq. (27):

$$P(Y > Z|M,R) = 1 - \Phi\left(\frac{\ln(z)-\mu_{erg}(M,R)}{\sigma_{erg}(M,R)}\right) \tag{27}$$

in which Y is the ground motion parameter of interest, $P(Y > Z|M,R)$ is the conditional probability that Y is larger than Z for a given combination of relevant parameters as magnitude M, distance R and others, and Φ is the cumulative distribution function of a standard normal distribution.

If a partially non-ergodic GMM for the site term is used, then the only change for the reference PSHA calculation is to use the single-station sigma for the standard deviation of the ground motion and add the constant shift to the median as shown in Eq. (28):

$$P(Y > Z|M,R) = 1 - \Phi\left(\frac{\ln(z)-(\mu_{erg}(M,R)+\delta S2S)}{\sigma_{SS}(M,R)}\right) \quad (28)$$

If a fully non-ergodic ground motion model is used, the coordinates of each source and site couple are needed to obtain the appropriate adjustment factors (which might need to be obtained by interpolation from pre-computed values). In this case, the probability of exceeding a given ground motion level Z is given by Eq. (29):

$$P(Y > Z|M,R) = 1 - \phi\left(\frac{\ln(z)-(\mu_{erg}(M,R)+\Delta\mu_{non-erg}(Lat_e,Lon_e,Lat_s,Lon_s,M))}{\sigma_{non-e} \quad (M,R)}\right) \quad (29)$$

in which Lat_e, Lon_e are the latitude and the longitude of the source and Lat_s, Lon_s are the latitude and longitude of the site. In general, hazard integral is computed site-by-site. The $\Delta\mu_{non-e}$ is the net adjustment factor for different source locations and magnitudes for the selected site. Finally, $\sigma_{non-erg}$ is the non-ergodic aleatory standard deviation.

In the current practice, to simplify the implementation in the PSHA software, the net adjustment terms used in each logic tree branch are computed before running the PSHA computation. In this way, few changes are needed in the PSHA software.

8.1.1. Multiple realization in logic tree for probabilistic seismic hazard calculations

Non-ergodic PSHA have been performed in several areas (e.g. Refs [24], [25]). In such applications, the epistemic uncertainty associated with the GMM is captured through the multiple realizations that sample the spatial distribution of the non-ergodic terms, which corresponds to a GMM logic tree with multiple branches. Each branch is a combination of an ergodic-based GMM and one realization of the net non-ergodic adjustment term, composed by the sum of all considered non-ergodic terms. In the logic tree sketch illustrated in Fig. 35, each branch of the GMM logic tree contains the net adjustment term which combines a source specific adjustment term, a site specific adjustment term and a cell specific path adjustment term. This node of the logic tree does not replace the epistemic uncertainty of the reference ergodic GMMs.

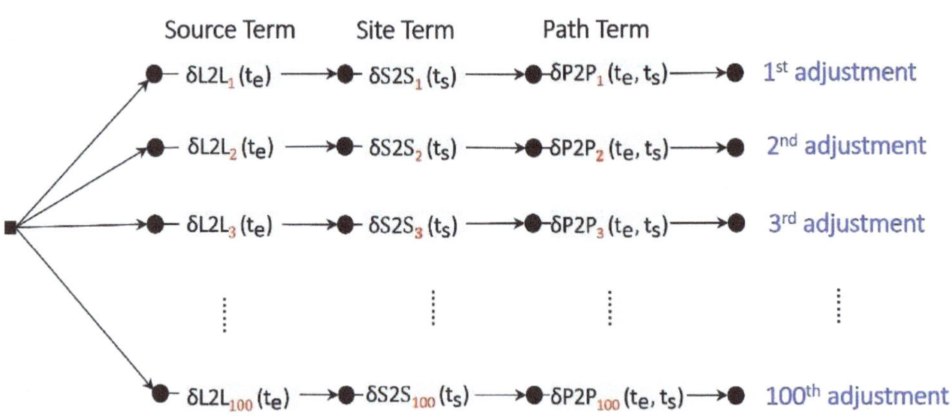

FIG. 35. Epistemic uncertainty inclusion by a logic tree with 100 realization branches of the non-ergodic adjustments (adapted from [25]).

Past studies have used 100 branches of the non-ergodic terms in the logic tree, but the hazard fractiles tend to be stable with 30 branches of the non-ergodic terms. For each logic tree component, the adjustment terms varying across the branches are sampled from their respective distributions. As such, for the site at which the non-ergodic PSHA is performed, uncertainty in the systematic site term is captured by sampling the site adjustment term distribution (100 samples of $\delta S2S_s$ in the example in Fig. 35, where *s* represents the target site). For the source adjustment term, 100 samples of spatially correlated values are taken (e.g. Ref. [24]). Such values are distributed according to a multivariate normal distribution with a covariance matrix whose elements describe the spatial correlation of the source terms across the considered region. Finally, for the path adjustment terms, multiple sets of cell specific linear path terms are sampled for different representative geometrical spreading, allowing to account for the correlation between the geometrical spreading and anelastic attenuation terms.

The whole set of realizations allow to properly sample the variance and the spatial correlation of the non-ergodic terms. An example is provided in Fig. 36, where the mean and the standard deviation of the adjustment factors per source grid point, computed over 100 realizations for two sites located in France, are shown.

In regions with less data, the median non-ergodic adjustment terms are close to zero, meaning that the median ground motion will be very close to the ergodic prediction, and the epistemic uncertainty in the median ground motion is large. As new data are collected and site, source and path specific terms become better constrained and the epistemic uncertainty in the median will be reduced. Note that the mean adjustments can be positive, negative, or close to zero. This relation between epistemic uncertainty and available data is not captured in ergodic models, which represents a limitation of hazard evaluations based on ergodic GMM.

8.1.2. Application of the non-ergodic ground motion models in deterministic seismic hazard calculations

The non-ergodic GMMs can also be used in deterministic seismic hazard analyses. The deterministic ground motion is computed using the median from the reference ergodic GMM plus the net non-ergodic term and using the reduced aleatory standard deviation. It is important to include a logic tree for the epistemic uncertainty in the non-ergodic terms for the deterministic approach. To account for the epistemic uncertainty, the weighted arithmetic mean of the deterministic ground motions from the different logic tree branches for the non-ergodic terms needs to be used, not the weighted geometric mean. With this approach, for a site without local data to constrain the non-ergodic terms, the weighted arithmetic mean for the 84th percentile ground motion will be similar to the ergodic 84th percentile ground motion.

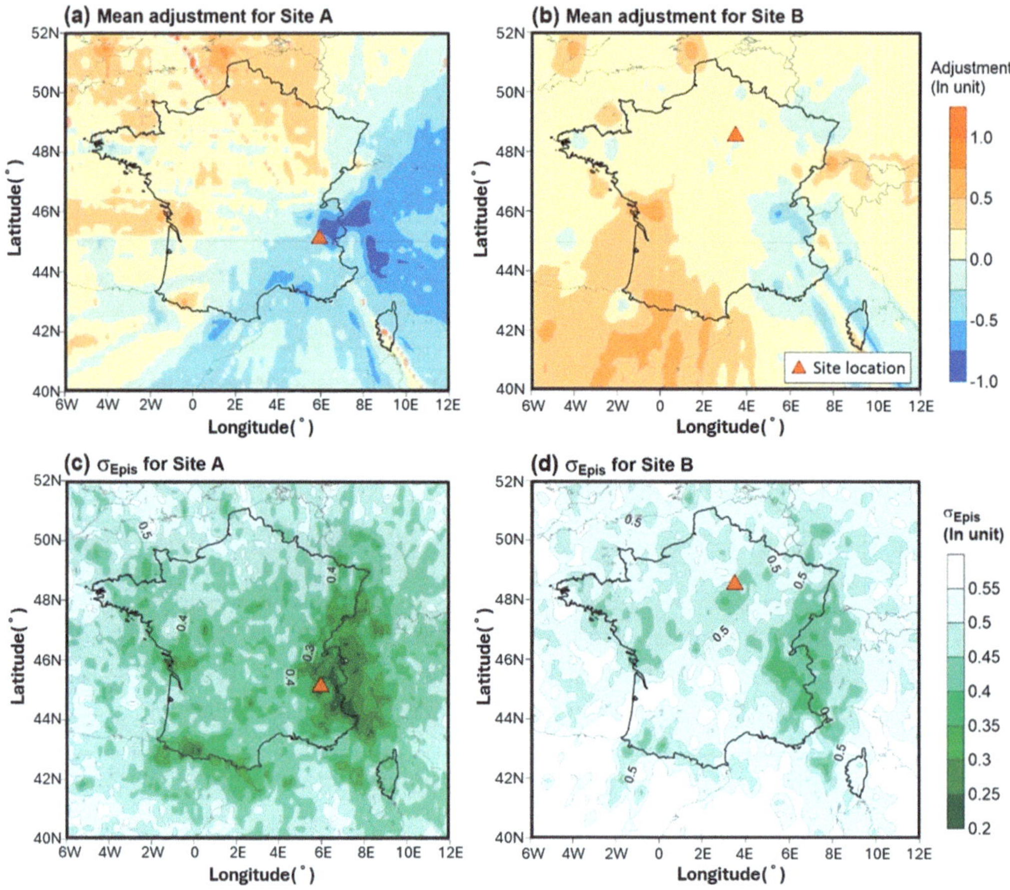

FIG. 36. (a, b) Mean and (c, d) standard deviation of 100 residuals per cell, which are defined as the differences between an ergodic ln(Y) and 100 non-ergodic ln(Y) at Site A (Longitude: 6.0 & Latitude:45.2) and Site B (Longitude: 3.52 & Latitude:48.5) at a selected frequency (adapted from [25]). Horizontal and vertical axes are the coordinates in longitude and latitude in degrees, respectively.

8.2.HAZARD SOFTWARE: KEY ASPECTS

Computing the hazard for 100 alternative GMMs can be a computational burden. More efficient methods can be developed using analytical approximations to model the effect of the epistemic uncertainty in the non-ergodic terms. One method is the Polynomial Chaos approach that uses a set of basis functions to approximate the change in the hazard due to a change in the median ground motion for normally distributed epistemic uncertainty in the non-ergodic terms. Reference [68] showed that the non-ergodic hazard for one site can be computed in about the time it takes to compute the hazard using 5 GMMs in a standard PSHA calculation.

The assumption of normally distributed epistemic uncertainty for the non-ergodic terms is appropriate for individual source locations, but it precludes combining the parts of a distributed seismicity source that are at the same distance for computational speed that is used in some PSHA programs. If parts of a distributed seismicity source zone that are at the same distance (rings around the site) are combined for computational speed, then the net distribution of the non-ergodic terms may be significantly different from a normal distribution. Methods for more efficient evaluations of the logic trees for non-ergodic GMMs are currently under development.

Computing probabilistic seismic hazard using non-ergodic GMMs will require modifications to the standard PSHA software. Before conducting non-ergodic PSHA calculations, verification of the modified PSHA software is essential.

8.3. EVALUATION OF NON-ERGODIC GROUND MOTION MODELS USING OBSERVATIONS

A recent IAEA publication provides state of the art practices and detailed technical elements related to the evaluation of PSHA based on observational data [69]. Reference [69] presents two levels of evaluation: the separate effect (elementary) evaluation that focusses on the logic tree elementary branches of the PSHA and the overall (integral) evaluation that focusses on the output hazard curves of the PSHA that combine multiple elementary branches.

The evaluation of GMMs is an important elementary step on the line of the integral PSHA evaluation. However, in the case of non-ergodic GMMs, as all recorded data are used to develop the non-ergodic ground motion model, it is difficult to evaluate the non-ergodic model using observational data. Nevertheless, there are two possible options:

- Option 1 is using recorded earthquakes that are not included in developing the non-ergodic GMM. While is not possible to have independent observed data to be used, alternative approaches such as utilization of data from other regions with similar tectonic context may be applicable but require adequate representation of uncertainties;
- Option 2 is using macroseismic intensity data (historical earthquakes).

8.3.1. Using empirical ground motions to evaluate the non-ergodic models

The concept of non-ergodic GMM evaluation is different of the one used for ergodic GMM evaluation. In fact, for a PSHA calculation using ergodic GMMs, several GMMs would be integrated in the logic tree with respective weights to constrain the epistemic uncertainty. The evaluation of ergodic GMMs with observation would allow to rank, score, or weight the different possible ergodic GMMs. The non-ergodic GMMs have already integrated recorded data. Then, the objective of the evaluation for non-ergodic GMM with independent observations is more about to assure (or verify) the reliability or performance of the non-ergodic GMM. The epistemic uncertainty for non-ergodic GMM is integrated by the logic tree using sampling methods as shows in precedents sections concerning the implementation.

An example for evaluating the performance of non-ergodic ground motion models in the Ridgecrest area is given in Ref. [70]. Using ergodic, partially non-ergodic (single-station sigma), and the fully non-ergodic GMMs to estimate the median ground motions from the earthquakes selected in the test set, the authors computed the root-mean-square errors (RMSE) with respect to the recorded values for the earthquakes which are not included in developing the GMMs. The RMSE is defined by the following equation:

$$RMSE = \sqrt{\frac{\sum_i (\hat{y}_i - y_i)^2}{N}} \quad (30)$$

where $\hat{y}_i$ and y_i are the observed and estimated median values for the ith ground motion recording. N is the total number of recordings for a given earthquake. Figure 37 shows the smoothed maps of the median peak ground acceleration (PGA) predicted by the different GMMs and the recorded PGA values for the earthquake with $M_w = 7.1$. The predicted median PGA values from the full non-ergodic and partially non-ergodic GMMs are more consistent with the recorded PGA than the ergodic model. There is a more significant variation of recorded PGA values in the southwest region, which is better retained by the non-ergodic as compared with ergodic GMM one. The predicted median values for full non ergodic and partially non ergodic GMMs are very similar with only minor differences. This similarity could result from the collocated nature of Ridgecrest earthquakes, which makes it difficult to separate site and path effects [70].

Figure 38 shows the histogram of residuals between recorded and median PGA predicted by the three GMMs. The variance of the residuals for the full GMM is slightly smaller than for the partially GMM, and both models show much smaller variances of residuals compared to the ergodic GMM of Ridgecrest earthquakes.

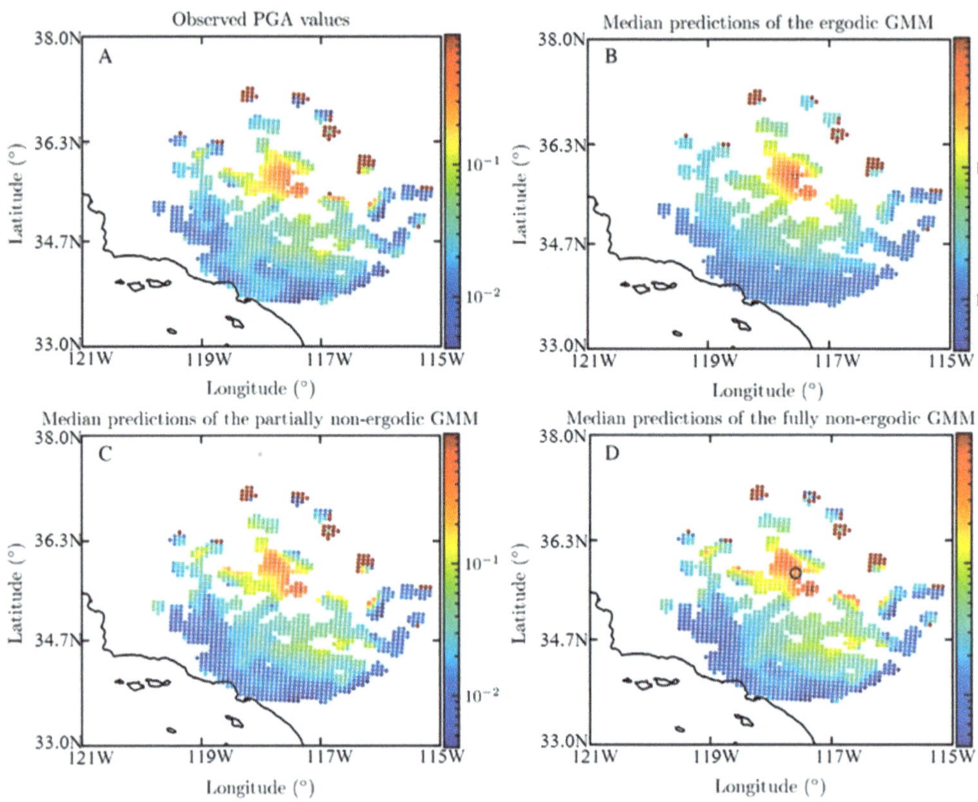

FIG. 37. Observed and predicted median values for PGA considering Mw = 7.1 using: (b) ergodic, (c) partially non-ergodic GMM, and (d) fully non-ergodic GMMs (adapted from [70]). Horizontal and vertical axes are the coordinates in longitude and latitude in degrees, respectively.

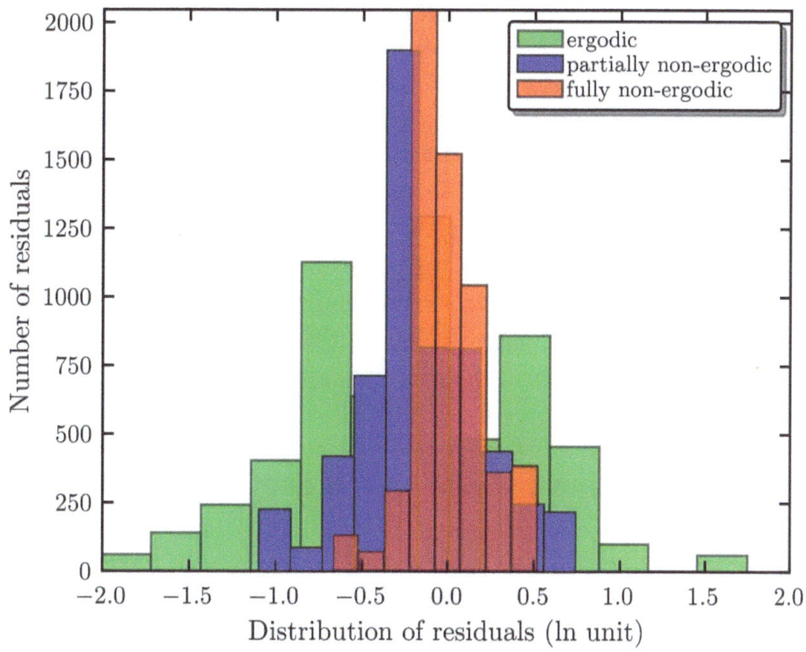

FIG. 38. Histograms of residuals (in ln units) for ergodic, SSS partially, and full non-ergodic GMMs, considering the earthquake scenario show in the precedent figure (adapted from [70]).

8.3.2. Using macroseismic data for evaluating non-ergodic models

The use of macroseismic data (i.e. macroseismic intensity) offers the possibility to test large magnitude earthquakes. Macroseismic data also provides denser spatial sampling of path effects than ground motion data. For regions of low seismicity, macroseismic data is a potentially useful source of information to evaluate non-ergodic GMMs; however, as the comparison is performed with GMM in terms of spectral acceleration, a transformation from macroseismic intensity to spectral acceleration (or the reverse) is need.

One approach is to convert the ground motions into macroseismic intensity estimates using ground motion intensity conversion equations (GMICE) such as in Ref. [71]. This has been a common approach, but Ref. [72] found that there can be a large bias in conversion for ground motion values not near the median for the scenario (large or small values of epsilon) if standard GMICE are used. As an alternative is to compare the spatial patterns of the residuals of the intensity data computed from intensity prediction equations with the spatial patterns of the ground motion residuals. For example, the two sets of residuals can be used to evaluate consistency of path effects from intensity data with the path effects from the non-ergodic GMM [73].

9. SUMMARY AND CONCLUSIONS

The main objective of this publication is to complement SSG-9 (Rev. 1) [2] with respect to the implementation of non-ergodic GMMs in the evaluation of seismic hazards in site evaluation for nuclear installations. As the practice of developing non-ergodic GMMs is relatively new, it is necessary to outline different ways of including non-ergodic terms in ground motion models using well established datasets of observed ground motion data, results for analytical modelling of site response, or full 3-D simulations from the source to the site. One important aspect of this publication is the presentation of consistent and unambiguous terminology for ground motion modelling with non-ergodic terms. Different levels of non-ergodicity in GMMs are defined, starting with the regionalization of ergodic ground motion models, inclusion of regional and site specific single-station sigma models, and finally the state of the art fully non-ergodic GMMs. This well defined terminology will help the Member States, particularly the readers in the nuclear practice, to comprehend and evaluate the differences in the current GMMs to be used in site evaluation projects.

A key aspect in the development of non-ergodic GMMs is the availability of ground motion data in the site region. Ongoing discussions focus on the spatial coverage (e.g. site specific or regional) and minimum amount of data needed to develop a non-ergodic GMM. These issues, particularly the data availability for nuclear installation sites, are elaborated in Section 3. In addition, the feasibility of developing a partially or fully non-ergodic GMMs for sites with limited data and the potential consequences (e.g. increase in the epistemic uncertainty) is discussed throughout the publication. These discussions underline two important issues: (1) the necessity of deploying a site specific seismic monitoring network at nuclear installation sites, and (2) the proper treatment of aleatory variability and epistemic uncertainties in PSHA. Most current applications of non-ergodic GMMs presented here show that even a limited amount of ground motion data, including data from small magnitude earthquakes, collected at the site is valuable in a non-ergodic approach to develop more accurate estimates of hazards for vibratory ground motion with an epistemic uncertainty range that captures the ‘true’ hazard (or the observations) at the site.

One of the sections of this publication is devoted to the support provided by ground motion simulations to the estimation of non-ergodic path and site terms. Improvements in the computational capabilities have increased the number of ground motion simulations in the last decade; however, including the data from ground motion simulations in ergodic GMMs has a limited potential to improve site specific seismic hazard estimates because site specific site, path, and source effects are not captured in ergodic GMMs, and the epistemic uncertainty range will be too narrow. Examples provided in this publication indicate the relevant types of ground motion simulation utilizing 3-D shear wave velocity models that may be accommodated by non-ergodic GMMs. Therefore, the ongoing practice in several Member States of estimating vibratory ground motion hazards by ground motion simulations will benefit from the examples provided in this publication.

An important message for the nuclear community is the possibility of developing non-ergodic GMMs for nuclear installations sites with no local ground motion data. For these cases, non-ergodic GMMs can still be used, but the epistemic uncertainty will be large. With simple calculations and sketches, it is demonstrated that the mean hazard will be the same in ergodic and non-ergodic PSHA for sites with no local ground motion data; however, the epistemic fractile range of the hazard for the ergodic GMM will be significantly underestimated compared to the non-ergodic GMM. If the fractiles of the hazard curves will be implemented in the safety assessment, then the difference will be significant when the non-ergodic approach is utilized. This difference needs to be understood and discussed in the review and assessment of safety analysis reports for nuclear installations. Combining the reduction in the aleatory variability with the change in the median ground motion for a given source-site combination, there can be either a net increase or decrease in the hazard at a given return period. The reduction in the aleatory variability will have an increasing effect for longer return periods, whereas at short return periods, the change in the median will affect the hazard at both short and long return periods.

As the non-ergodic PSHA practice is quite new; there are few PSHA software that can be used in PSHA calculations. Therefore, the user needs to be aware of the changes in the hazard calculation equations and software if a non-ergodic GMM is implemented in the hazard analysis. In addition, validation and verification of the hazard code for proper implementation of non-ergodic GMMs is particularly important in nuclear projects.

REFERENCES

[1] INTERNATIONAL ATOMIC ENERGY AGENCY, Site Evaluation for Nuclear Installations, IAEA Safety Standards Series No. SSR-1, IAEA, Vienna (2019).

[2] INTERNATIONAL ATOMIC ENERGY AGENCY, Seismic Hazards in Site Evaluation for Nuclear Installations, IAEA Safety Standards Series No. SSG-9 (Rev. 1), IAEA, Vienna (2022).

[3] ANCHETA, T.D., et al., NGA-West2 database, Earthquake Spectra **30**(3) (2014) 989-1005, https://doi.org/10.1193/070913eqs197m.

[4] GOULET, C.A., et al. PEER NGA-east database, Earthquake Spectra **37** (2021) 1331-1353, https://doi.org/10.1177/87552930211015695.

[5] MAZZONI, S., et al. Relational database used for ground-motion model development in the NGA-Sub project, Earthquake Spectra **38**(2) (2022) 1529-1548, https://doi.org/10.1177/87552930211055204.

[6] LANZANO, G., et al. The pan-European Engineering Strong Motion (ESM) flatfile: compilation criteria and data statistics, Bulletin of Earthquake Engineering **17** (2019) 561-582, https://doi.org/10.1007/s10518-018-0480-z.

[7] INTERNATIONAL ATOMIC ENERGY AGENCY, Seismic Hazard Assessment in Site Evaluation for Nuclear Installations: Ground Motion Prediction Equations and Site Response, IAEA-TECDOC-1796, IAEA, Vienna (2016).

[8] BALTAY, A. S., BOATWRIGHT, J., Ground-motion observations of the 2014 South Napa earthquake, Seismological Research Letters **86**(2A) (2015) 355-360, https://doi.org/10.1785/0220140232.

[9] GRAIZER, V., KALKAN, E., Summary of the GK15 ground-motion prediction equation for horizontal PGA and 5% damped PSA from shallow crustal continental earthquakes, Bulletin of the Seismological Society of America **106**(2) (2016) 687-707, https://doi.org/10.1785/0120150194.

[10] ABRAHAMSON, N., KUEHN, N., WALLING, M., LANDWEHR, N., Probabilistic seismic hazard analysis in California using nonergodic ground-motion models, Bulletin of the Seismological Society of America **109** (2019) 1235-1249, https://doi.org/10.1785/0120190030.

[11] ANDERSON, J.G., BRUNE, J. N., Probabilistic seismic hazard analysis without the ergodic assumption, Seismological Research Letters **70** (1999) 19-28, https://doi.org/10.1785/gssrl.70.1.19.

[12] BOMMER, J.J., ABRAHAMSON, N.A., Why do modern probabilistic seismic-hazard analyses often lead to increased hazard estimates? Bulletin of the Seismological Society of America **96** (2006) 1967-1977, https://doi.org/10.1785/0120060043.

[13] ATIK, L.A., et al., The variability of ground-motion prediction models and its components, Seismological Research Letters **81** (2010) 794-801, https://doi.org/10.1785/gssrl.81.5.794.

[14] BOZORGNIA, Y. et al., NGA-West2 research project, Earthquake Spectra **30**(3) (2014) 973-987, https://doi.org/10.1193/072113EQS209M.

[15] BOZORGNIA, Y. et al., NGA-Subduction research program, Earthquake Spectra **38**(2) (2022) 783-798, https://doi.org/10.1177/87552930211056081.

[16] ABRAHAMSON, N.A., GULERCE, Z., Summary of the Abrahamson and Gulerce NGA-SUB ground-motion model for subduction earthquakes, Earthquake Spectra **38** (2022) 2638-2681, https://doi.org/10.1177/87552930221114374.

[17] GULERCE, Z., KARGOIĞLU, B., ABRAHAMSON, N.A., Turkey-adjusted NGA-W1 horizontal ground motion prediction models, Earthquake Spectra **32** (2016) 75-100, https://doi.org/10.1193/022714EQS034M.

[18] SUNG, C.H., ABRAHAMSON, N.A., LACOUR, M., Methodology for including path effects due to 3D velocity structure in nonergodic ground-motion models, Bulletin of the Seismological Society of America **113** (2023) 2144-2163, https://doi.org/10.1785/0120220252.

[19] LAVRENTIADIS, G., ABRAHAMSON, N. A., KUEHN, N. M., A non-ergodic effective amplitude ground-motion model for California, Bulletin of Earthquake Engineering 21 (2023) 5233-5264, https://doi.org/10.1007/s10518-021-01206-w.

[20] AKKAR, S. et al., Reference database for seismic ground-motion in Europe (RESORCE), Bulletin of Earthquake Engineering **12** (2014) 311-339, https://doi.org/10.1007/s10518-013-9506-8.

[21] PAOLUCCI, R. et al., "Record processing in ITACA, the new Italian strong-motion database", Earthquake Data in Engineering Seismology: Predictive Models, Data Management and Networks, 14 (AKKAR, S., GULKAN, P., VAN ECK, T., Eds) Springer, Dordrecht, https://doi.org/10.1007/978-94-007-0152-6_8.

[22] TRAVERSA, P. et al., RESIF RAP and RLBP dataset of earthquake ground motion in mainland France, Seismological Research Letters **91**(4) (2020) 2409-2424, https://doi.org/10.1785/0220190367.

[23] MACEDO, J., LIU, C., A nonergodic ground motion model for Chile, Bulletin of the Seismological Society of America **112** (2022) 2542-2561, https://doi.org/10.1785/0120210334.

[24] LANDWEHR, N., KUEHN, N. M., SCHEFFER, T., ABRAHAMSON, N.A., A nonergodic ground-motion model for California with spatially varying coefficients, Bulletin of the Seismological Society of America **106**(6) (2016) 2574-2583, https://doi.org/10.1785/0120160118.

[25] SUNG, C-H. et al., A non-ergodic ground-motion model of Fourier amplitude spectra for France, Bulletin of Earthquake Engineering **21**(11) (2023) 5293-5317, https://doi.org/10.1007/s10518-022-01403-1.

[26] RODRIGUEZ-MAREK, A. et al., A model for single-station standard deviation using data from various tectonic regions, Bulletin of the Seismological Society of America **103**(6) (2013) 3149-3163, https://doi.org/10.1785/0120130030.

[27] FACCIOLI, E., PAOLUCCI, R., VANINI, M., Evaluation of probabilistic site-specific seismic-hazard methods and associated uncertainties with applications in the Po Plain, northern Italy, Bulletin of the Seismological Society of America **105** (2015) 2787-2807, https://doi.org/10.1785/0120150051.

[28] KOTHA, S. R., BINDI, D., COTTON, F., From ergodic to region-and site-specific probabilistic seismic hazard assessment: Method development and application at European and Middle Eastern sites, Earthquake Spectra **33**(4) (2017) 1433-1453, https://doi.org/10.1193/081016eqs130m.

[29] PACIFIC GAS AND ELECTRIC COMPANY, Seismic Source Characterization for the Diablo Canyon Power Plant, San Luis Obispo County, California: Report on the Results of SSHAC Level 3 Study (Rev. A) (2015).

[30] FRANKEL, A., WIRTH, E., MARAFI, N., VIDALE, J., STEPHENSON, W., Broadband synthetic seismograms for magnitude 9 earthquakes on the Cascadia megathrust based on 3D simulations and stochastic synthetics, Part 1: Methodology and overall results, Bulletin of the Seismological Society of America **108**(5A) (2018) 2347-2369, https://doi.org/10.1785/0120180034.

[31] PAOLUCCI, R., MAZZIERI, I., SMERZINI, C., Anatomy of strong ground motion: near-source records and three-dimensional physics-based numerical simulations of the Mw 6.0 2012 May 29 Po Plain earthquake, Italy, Geophysical Supplements to the Monthly Notices of the Royal Astronomical Society **203**(3) (2015) 2001-2020, https://doi.org/10.1093/gji/ggv405.

[32] BOORE, D. M., Simulation of ground motion using the stochastic method, Pure and Applied Geophysics **160** (2003) 635-676, https://doi.org/10.1007/PL00012553.

[33] INTERNATIONAL ATOMIC ENERGY AGENCY, IAEA Nuclear Safety and Security Glossary: Terminology Used in Nuclear Safety, Nuclear Security, Radiation Protection and Emergency Preparedness and Response, 2022 (Interim) Edition, IAEA, Vienna (2022), https://doi.org/10.61092/iaea.rrxi-t56z.

[34] MARZOCCHI, W., JORDAN, T. H., Testing for ontological errors in probabilistic forecasting models of natural systems, Proceedings of the National Academy of Sciences **111(**33) (2014) 11973-11978, https://doi.org/10.1073/pnas.1410183111.

[35] MILNER, K.R., et al., Toward physics-based nonergodic PSHA: A prototype fully deterministic seismic hazard model for southern California, Bulletin of the Seismological Society of America **111**(2) (2021) 898-915, https://doi.org/10.1785/0120200216.

[36] WANG, F., JORDAN, T. H., Comparison of probabilistic seismic-hazard models using averaging-based factorization, Bulletin of the Seismological Society of America **104** (2014) 1230-1257, https://doi.org/10.1785/0120130263.

[37] LIOU, I., ABRAHAMSON, N. A., Aleatory variability and epistemic uncertainty for ground-motion models in PSHA, submitted to Bulletin of the Seismological Society of America (2025).

[38] LIN, P-S., et al., Repeatable source, site, and path effects on the standard deviation for empirical ground-motion prediction models, Bulletin of the Seismological Society of America **101**(5) (2011) 2281-2295, https://doi.org/10.1785/0120090312.

[39] ATKINSON, G.M., BOMMER, J.J., ABRAHAMSON, N.A., Alternative approaches to modelling epistemic uncertainty in ground motions in probabilistic seismic-hazard analysis, Seismological Research Letters **85** (2014) 1141-1144, https://doi.org/10.1785/0220140120.

[40] GEOPENTECH, Southwestern United States Ground-Motion Characterization, SSHAC Level 3 - Technical report (Rev 2) (2015).

[41] RAI, M., RODRIGUEZ-MAREK, A., CHIOU, B.S., Empirical terrain-based topographic modification factors for use in ground motion prediction, Earthquake Spectra **33** (2017) 157-177, https://doi.org/10.1193/071015eqs111m.

[42] ABRAHAMSON, N. A., HOLLENBACK, J. C., "Application of single-station sigma ground motion prediction equations in practice" (Proc. of the 15th World Conference on Earthquake Engineering), Lisbon, Portugal (2012).

[43] STAFFORD, P.J., RODRIGUEZ-MAREK, A., EDWARDS, B., KRUIVER, P.P., BOMMER, J.J., Scenario dependence of linear site-effect factors for short-period response spectral ordinates, Bulletin of the Seismological Society of America **107**(6) (2017) 2859-2872, https://doi.org/10.1785/0120170084.

[44] LI, M., ABRAHAMSON, N., RATHJE, E., Issues for ground-motion models using small-magnitude events, submitted to The Seismic Record (2025).

[45] SWISSNUCLEAR, Probabilistic Seismic Hazard Analysis for Swiss Nuclear Power Plant Sites - PEGASOS Refinement Project (Rev. 1, Final Report). Vol. **1-5**. (2014) URL: https://swissnuclear.ch/en/links-und-downloads/

[46] SENNA, S., et al., Modelling of the subsurface structure from the seismic bedrock to the ground surface for a broadband strong motion evaluation, Journal of Disaster Research **8** (2013) 889-903, https://doi.org/10.20965/jdr.2013.p0889.

[47] RODRIGUEZ-MAREK, A., et al., SSHAC Level 2 Site Response Study for Two Example US Sites, NRC Headquarters, Rockville, Maryland (2020).

[48] STEWART, J.P., AFSHARI, K., GOULET, C.A., Non-ergodic site response in seismic hazard analysis, Earthquake Spectra **33** (2017) 1385-1414, https://doi.org/10.1193/081716eqs135m.

[49] LIOU, I., ABRAHAMSON, N.A., Approaches to soil hazard in partially non-ergodic ground-motion models, submitted to Bulletin of the Seismological Society of America (2025).

[50] RODRIGUEZ-MAREK, A., MONTALVA, G. A., COTTON, F., BONILLA, F., Analysis of single-station standard deviation using the KiK-net data, Bulletin of the Seismological Society of America **101**(3) (2011) 1242-1258, https://doi.org/10.1785/0120100252.

[51] DAWOOD, H. M., RODRIGUEZ-MAREK, A., A method for including path effects in ground-motion prediction equations: An example using the Mw 9.0 Tohoku earthquake aftershocks, Bulletin of the Seismological Society of America **103**(2B) (2013) 1360-1372, https://doi.org/10.1785/0120120125.

[52] KUEHN, N. M., ABRAHAMSON, N. A., WALLING, M. A., Incorporating nonergodic path effects into the NGA-West2 ground-motion prediction equations, Bulletin of the Seismological Society of America **109** (2019) 575-585, https://doi.org/10.1785/0120180260.

[53] VILLANI, M., ABRAHAMSON, N. A., Repeatable site and path effects on the ground-motion sigma based on empirical data from southern California and simulated waveforms from the CyberShake platform, Bulletin of the Seismological Society of America **105** (2015) 2681-2695, https://doi.org/10.1785/0120140359.

[54] MENG, X., GOULET, C. A., Lessons learned from applying varying coefficient model to controlled simulation datasets, Bulletin of Earthquake Engineering **21** (2023) 5151–5174, https://doi.org/10.1007/s10518-022-01512-x.

[55] MAECHLING, P. J., SILVA, F., CALLAGHAN, S., JORDAN T.H., SCEC Broadband Platform: System Architecture and Software Implementation, Seismological Research Letters **86** (2015) 27-38, https://doi.org/10.1785/0220140125.

[56] GRAVES, R., PITARKA, A., Refinements to the Graves and Pitarka (2010) broadband ground-motion simulation method, Seismological Research Letters **86** (2015) 75-80, https://doi.org/10.1785/0220140101.

[57] OLSEN, K., TAKEDATSU, R., The SDSU broadband ground-motion generation module BBtoolbox version 1.5, Seismological Research Letters **86** (2015) 81-88, https://doi.org/10.1785/0220140102.

[58] CREMPIEN, J. G. F., ARCHULETA, R. J., UCSB method for simulation of broadband ground motion from kinematic earthquake sources, Seismological Research Letters **86** (2015) 61-67.

[59] IWAKI, A., MAEDA, T., MORIKAWA, N., MIYAKE, H., FUJIWARA, H., Validation of the recipe for broadband ground-motion simulations of Japanese crustal earthquakes, Bulletin of the Seismological Society of America **106**(5) (2016) 2214-2232, https://doi.org/10.1785/0120150304.

[60] ATKINSON, G.M., ASSATOURIANS, K., Implementation and validation of EXSIM (a stochastic finite-fault ground-motion simulation algorithm) on the SCEC broadband platform, Seismological Research Letters **86** (2015) 48-60, https://doi.org/10.1785/0220140097.

[61] SONG, S.G., Developing a generalized pseudo-dynamic source model of Mw 6.5–7.0 to simulate strong ground motions, Geophysical Journal International **204** (2016) 1254-1265, https://doi.org/10.1093/gji/ggv521.

[62] PITARKA, A., GRAVES, R., IRIKURA, K., MIYAKE, H., RODGERS, A., "Performance of Irikura recipe rupture model generator in earthquake ground motion simulations with Graves and Pitarka hybrid approach", Best Practices in Physics-based Fault Rupture Models for Seismic Hazard Assessment of Nuclear Installations (2018), https://doi.org/10.1007/978-3-319-72709-7.

[63] MCCALLEN, D., et al., EQSIM—A multidisciplinary framework for fault-to-structure earthquake simulations on exascale computers part I: Computational models and workflow, Earthquake Spectra **37**(2) (2021) 707-735, https://doi.org/10.1177/8755293020970982.

[64] SHAHJOUEI, A., PEZESHK, S., Synthetic seismograms using a hybrid broadband ground-motion simulation approach: Application to central and eastern United States, Bulletin of the Seismological Society of America **105** (2015) 686-705, https://doi.org/10.1785/0120140219.

[65] https://www.scec.org/software/cybershake

[66] SUNG, C.H., ABRAHAMSON, N. A., Partially nonergodic ground-motion model for Cascadia interface earthquakes, Bulletin of the Seismological Society of America **112** (2022) 2520-2541, https://doi.org/10.1785/0120210330.

[67] MAEDA, T., IWAKI, A., MORIKAWA, N., AOI, S., FUJIWARA, H., Seismic-hazard analysis of long-period ground motion of megathrust earthquakes in the Nankai trough based on 3D finite-difference simulation, Seismological Research Letters **87**(6) (2016) 1265-1273, https://doi.org/10.1785/0220160093.

[68] LACOUR, M., ABRAHAMSON, N. A., Efficient propagation of epistemic uncertainty in the median ground-motion model in probabilistic hazard calculations, Bulletin of the Seismological Society of America **109** (2019) 2063-2072, https://doi.org/10.1785/0120180327.

[69] INTERNATIONAL ATOMIC ENERGY AGENCY, Evaluation of Probabilistic Seismic Hazard Analysis (PSHA) Based on Observational Data, IAEA-TECDOC-2047, IAEA, Vienna (2024).

[70] LIU, C., MACEDO, J., KUEHN, N., Spatial correlation of systematic effects of non-ergodic ground motion models in the Ridgecrest area, Bulletin of Earthquake Engineering **21** (2023) 5319–5345, https://doi.org/10.1007/s10518-022-01441-9.

[71] WORDEN, C. B., GERSTENBERGER, M.C., RHOADES, D.A., WALD, D.J., Probabilistic relationships between ground-motion parameters and modified Mercalli intensity in California, Bulletin of the Seismological Society of America **102**(1) (2012) 204-221, https://doi.org/10.1785/0120110156.

[72] GALLAHUE, M., ABRAHAMSON, N.A., New methodology for unbiased ground-motion intensity conversion equations, Bulletin of the Seismological Society of America **113** (2023) 1133-1151, https://doi.org/10.1785/0120220224.

[73] ABRAHAMSON, N. A., SUNG, C-H., "Non-Ergodic Ground-Motion Models Based on A Combined Set of Empirical Ground-Motion Data, 3-D Simulations, and Macro-Seismic Intensity Data" (Proc. of Anniversary Workshop in Commemoration of the 1999 Chi-Chi and 2022 Chihshang Earthquakes, Taipei (2023).

ANNEX I. PRELIMINARY NON-ERGODIC GROUND MOTION MODEL FOR CZECH REPUBLIC

The geological bedrock of majority of the Czech Republic territory consists of the Bohemian Massif, which is a remnant of the collisional Variscan Orogeny. The Bohemian Massif is also an important seismotectonic domain of Central Europe, which is adjacent to the more seismically active Alpine-Carpathian chain in the south, southeast and east. Therefore, earthquakes occurring in the Eastern Alps contribute significantly to the seismic hazards for the sites at the Bohemian Massif. Macroseismic studies of the historical earthquakes in the Eastern Alps and Western Carpathians have pointed out an elongation of isoseismals in the northwest direction in the pre-instrumental and early instrumental periods. This elongation is attributed to an attenuation anomaly in the central Europe.

The first study on this attenuation phenomenon was based on the distribution of macroseismic observations in Bohemia and Moravia from the pre-instrumental Eastern Alpine earthquakes occurred in 1927, 1938, and 1939 [I–1]. The study concluded that the macroseismic fields were elongated to the N-NW, reaching as far as the city of Dresden. The same attenuation phenomenon was observed in the macroseismic fields of the earthquakes from other source zones, both within the Bohemian Massif and outside, e.g. in the Western Carpathians. The earthquake occurred near Žilina in 1858, the macroseismic field of which was captured on a coloured isoseismal map immediately after the earthquake, may be considered as a good example [I–2]. The macroseismic contours of the Žilina earthquake extended significantly in the NW direction and reached as far as 270 km to Jelenia Góra in Poland. An example of the elongated isoseismals of an earthquake occurred on April 16, 1972 with an epicentre near Seebenstein (Lower Austria) on the Vienna Basin is shown in Fig. I–1.

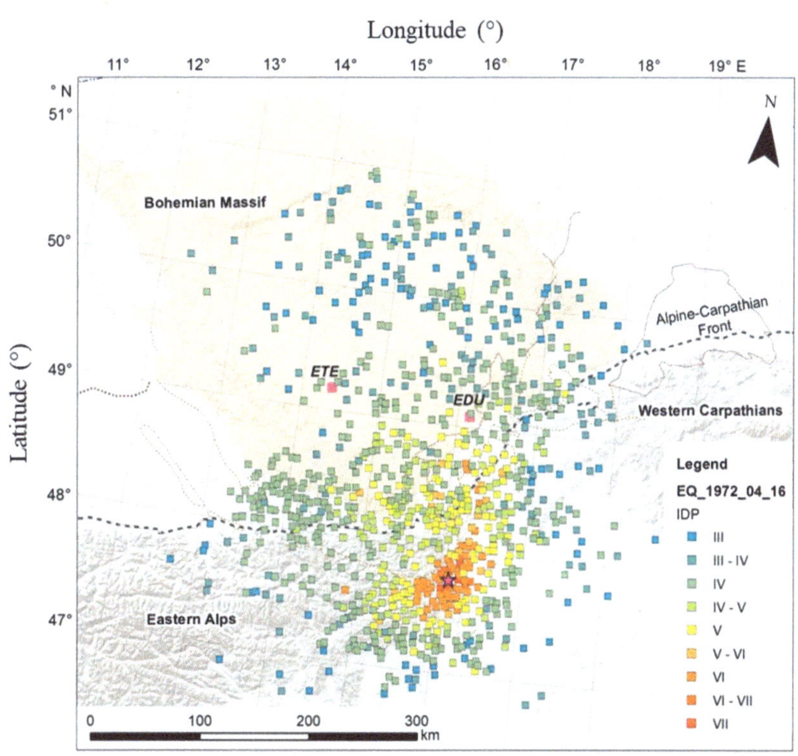

FIG. I–1. Distribution of the macroseismic intensity data points (IDP) of the April 16, 1972 earthquake showing the extension of the macroseismic field from the Alpine source area to the Bohemian Massif (yellow-brown shading of surface indicates the pre-Tertiary rocks). The epicentre is marked with a red star (data taken from Ref. [I–3]). Horizontal and vertical axes are the coordinates in longitude and latitude in degrees, respectively.

The irregular attenuation of seismic waves in Central Europe were also observed in the recorded ground motions of two moderate-size earthquakes in the Vienna Basin, which occurred in September and October of 2013 [I–4]. The peak amplitudes of both velocity and acceleration time histories to the northwest of the epicentres were found to be higher than the peak amplitudes recorded in other directions. This difference is possibly associated with the propagation of the S-waves. New studies have confirmed the existence of the attenuation irregularity as known as the mid-European anomaly in the seismological community, but the underlying cause of the irregularity is not yet fully recognized.

The mid-European attenuation anomaly results in a significant path-dependency of the ground shaking, which may be considered in the seismic hazard assessment by using a non-ergodic ground motion model. The Czech Republic operates two nuclear power plants at the Temelín (2 units, each with the capacity of 1055 MWe) and Dukovany (4 units, each with the capacity of 510 MWe) sites. There are also plans for the construction of new units at these existing sites and advanced reactors (e.g. small modular reactors) in new sites. The most current probabilistic seismic hazard assessment (PSHA) for the Temelín and Dukovany sites was conducted in 2022 [I–5] and the existence of the path-dependency of the ground shaking was considered as a significant source of uncertainty in the PSHA. It is not possible to fully address this dependency by using global ergodic ground motion models, and utilizing the ergodic GMMs in PSHA may lead to the underestimation of seismic hazard. On the other hand, developing a non-ergodic ground motion model may be difficult in areas with low seismicity, such as the Bohemian Massif, due to data scarcity.

This annex summarizes the first attempt to develop a preliminary non-ergodic ground motion model for Czech Republic, as an example of developing non-ergodic GMMs for regions with limited available data.

I-1. COMPILATION OF THE DATABASE AND DEVELOPMENT OF REFERENCE ERGODIC GROUND MOTION MODEL

Earthquakes in the Bohemian Massif have been instrumentally recorded since November 28, 1908, when the first seismic station in Cheb was put into operation. Electronic bulletins of seismic events (with onset times and maximal amplitudes) are available since January of 1976. Currently, there are 20 permanent seismic stations in the Czech Republic operated by the Czech Regional Seismic Network, CRSN, and many other stations are available in local networks. The coverage of the region by seismic stations increased significantly during the Alp Array international experiment.

For the preliminary ground motion model, ground motion recordings from 13 earthquakes (red stars in Fig. I–2) were compiled. Four of them were induced seismicity events located in mining areas such as the Upper Silesian coal basin (#7 and 11 in Fig. I–2) and Lubin copper mine (#3 and 10). Two events (#8 and 9) are from West Bohemia, which is a well known area for occurrence of seismic swarms. Furthermore, events from the Eastern Alps (#1), the Vienna Basin (#5), the Western Carpathians (#12), the vicinity of Leipzig (#4), the foothills of the Ore Mountains (#2) and the Tyrolean Unterland (#13) were included.

The number of recordings per event varies significantly as only the recordings from broadband and short-period recording stations with high sampling frequencies (more than 80 Hz) were included in the dataset (Table I–1). Although the number of useful stations was significantly reduced with this limitation, collected data can be used for a large range of frequency (1–35 Hz). This frequency range cover the most critical ground motion frequencies for nuclear power plants. Selected seismograms were visually inspected to eliminate recordings with high seismic noise or non-standard errors. For consistency, all seismograms were corrected for frequency characteristics of the sensor and filtered using a band-pass filter of 1–35 Hz.

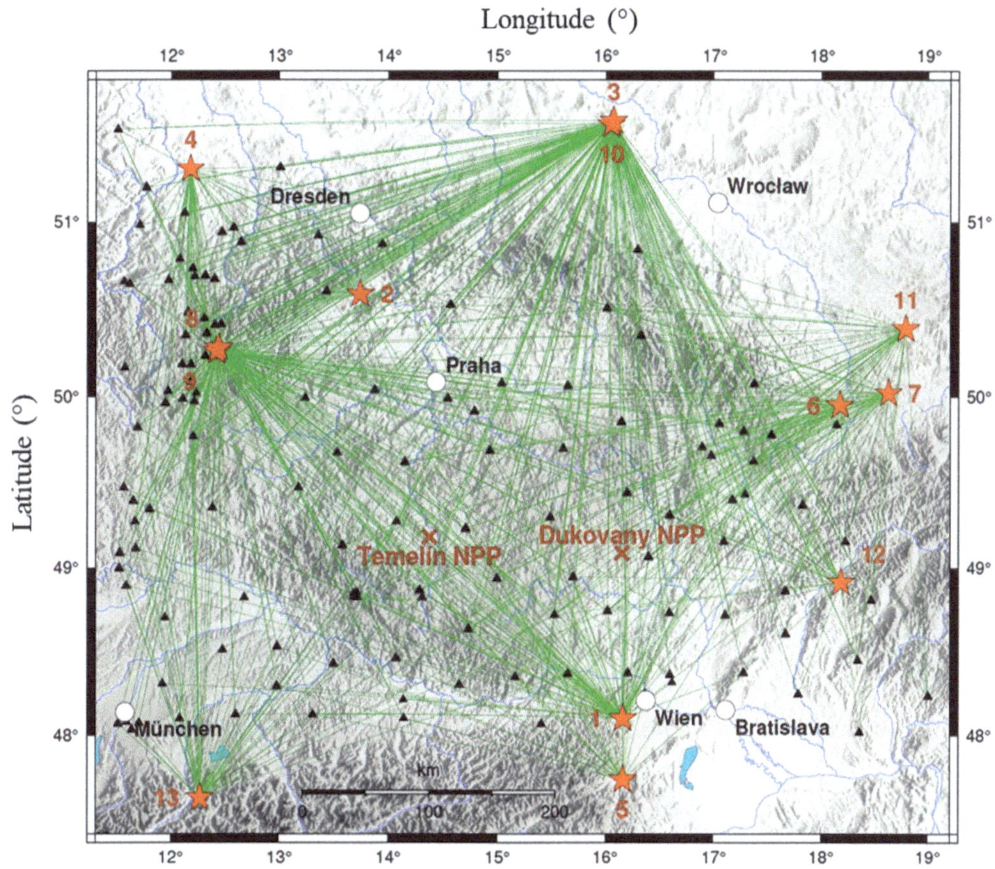

FIG. I–2. Locations of earthquakes (stars) and stations (triangles) in the compiled dataset. The green lines show the sampling of the ray paths. Earthquake numbers (1-13) are consistent with Table I–1. Horizontal and vertical axes are the coordinates in longitude and latitude in degrees, respectively.

TABLE I–1. EARTHQUAKES COMPILED IN THE DATASET AND NUMBER OF RECORDINGS FORM THE EARTHQUAKES.

Event number	Date	Moment magnitude	Hypocentre			Number of recordings	Hypocentral Distance of the Recordings (km)	
			Longitude	Latitude	Depth (km)		Min	Max
1	2016-04-25	4.2	16.16	48.10	5	84	42	385
2	2017-02-17	2.3	13.74	50.59	3	11	22	150
3	2017-04-08	4.7	16.07	51.57	10	95	82	396
4	2017-04-29	3.1	12.18	51.30	10	49	28	272
5	2017-11-10	3.2	16.16	47.73	5	28	60	274
6	2017-12-10	3.6	18.17	49.95	10	42	13	292
7	2018-05-05	4.3	18.63	50.02	10	43	41	388
8	2018-05-14	3.7	12.44	50.28	5	98	5	292
9	2018-05-21	4.2	12.42	50.26	4	102	4	371
10	2018-07-20	4.8	16.05	51.55	5	102	81	394
11	2018-08-11	4.1	18.79	50.39	10	39	77	401
12	2018-08-17	3.1	18.18	48.91	5	32	24	262
13	2019-10-22	4.3	12.27	47.62	3	62	55	378

Thc final dataset used for developing the reference ergodic and non-ergodic GMMs consists of 786 recordings from three earthquakes recorded at 152 stations between 2016 and 2019. The magnitudes of the events range from M_w=2.3 to M_w=4.8 and hypocentral distances of the recordings range from 4 to 400 km. The distribution of the data in magnitude and distance space is shown in Fig. I–3. For these recordings, peak ground acceleration (PGA) and the peak ground velocity (PGV) values for two orthogonal horizontal components were added to the database and the geometric mean of the horizontal components was used in GMM development.

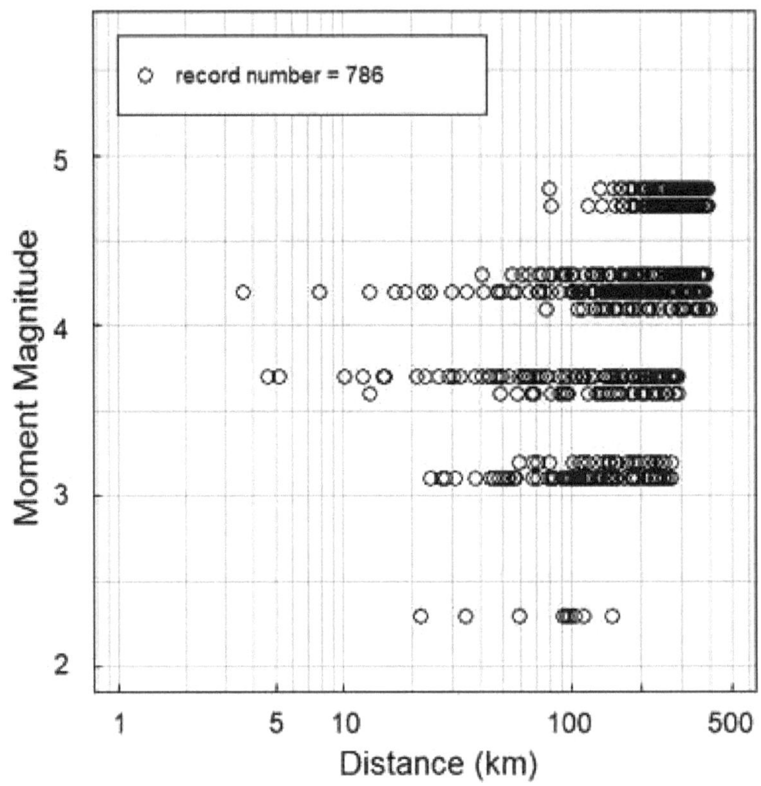

FIG. I–3. Magnitude and hypocentral distance of the recordings.

Using the compiled dataset, the reference ergodic GMM for PGA and PGV with a simple functional form was developed as shown in Equation (I–1):

$$ln(Y_{es}) = c_o + c_1 M_w + c_2 ln(R_{Hypo} + 1) + c_3 R_{Hypo} + \delta B_e + \delta W_{es} \qquad \text{(I–1)}$$

where M_w is the moment magnitude, R_{Hypo} is the hypocentral distance in km, and δB_e and δW_{es} are the between-event and within-event residuals with standard deviations of τ and ϕ, respectively. The site conditions of the recording stations were not available, so the reference ergodic GMM does not include a site term.

The between-event and within-event standard deviation values of the reference ergodic model is provided in Table I–2. The within-event standard deviations for PGA and PGV are larger than the values expected for ergodic GMMs because the site terms were not included in the functional form. As the ergodic site terms were not included, the non-ergodic GMM will account for the median site scaling in the non-ergodic site terms. The between-event standard deviations values are also large (1.0 for PGA and 0.74 for PGV in ln units) which may result from the uncertainty in moment magnitude values estimated from the local magnitude scales by the magnitude conversion equation.

TABLE I-2. RESIDUALS AND COMPONENTS OF THE STANDARD DEVIATION (IN LN UNITS) FOR THE ERGODIC AND NON-ERGODIC GMMS. ($\delta P2P_Q$, $\delta P2P_V$, δWSP ARE DEFINED IN SECTION 6.1 AND THE OTHER TERMS ARE GIVEN IN TABLE 1 (SECTION 4)).

Residual Type	Standard Deviation	Type of the GMM	Standard Deviation Value	
			ln (PGA)	ln (PGV)
δB	τ	Ergodic	1.001	0.741
δW	ϕ	Ergodic	0.789	0.682
δS2S	ϕ_{S2S} (combined regional and uncorrelated terms)	Non-ergodic	0.619	0.501
δWS	ϕ_{SS}	partially non-ergodic	0.488	0.463
$\delta P2P_Q$	$\phi_{P2P\text{-}Q}$	non-ergodic	0.281	0.247
$\delta P2P_V$	$\phi_{P2P\text{-}V}$	non-ergodic	0.340	0.358
δWSP	ϕ_{SP}	non-ergodic	0.208	0.182

I-2. PRELIMINARY NON-ERGODIC GROUND MOTION MODEL

The preliminary non-ergodic GMM is developed by following the methods described in Section 5 and 6. Both non-ergodic site and path terms are included in the functional form. The non-ergodic source terms are not included, due to the possible errors in moment magnitude estimates and the limited number of earthquakes in the dataset. The functional form of the non-ergodic GMM is given by Equation (I–2):

$$\delta W_{es} = a_0 + \delta S2S_{reg}(x_s) + \delta S2S_{ncor,s} + \delta P2P_Q(x_e, x_s) + \delta P2P_V(x_e, x_s) + \delta WSP \quad \text{(I-2)}$$

where x_s and x_e are the coordinates of the site and earthquake, respectively, $\delta S2S_{reg}(x_s)$ is a regional site term that is spatially correlated, $\delta S2S_{ncor,s}$ is the remaining site term that is not spatially correlated, $\delta P2P_Q(x_e, x_s)$ is the path term related to the regional variations in crustal attenuation (using the cell approach), and $\delta P2P_V(x_e, x_s)$ is the path term related to the 3-D velocity structure. The δWSP is the remaining within-event single path residual.

The components of the standard deviation for PGA and PGV are listed in Table I–2. The standard deviations of the systematic site terms (ϕ_{S2S}) are large because they include the median site scaling in addition to the variability of the site specific amplification compared to the site scaling in the GMM. The single-station within-event standard deviations, ϕ_{SS}, are consistent with the ϕ_{SS} values found for other data sets: previous studies reported ϕ_{SS}=0.48 ln units for California data and ϕ_{SS}=0.46 - 0.50 ln units for European data [I–6]. The standard deviations of the two path terms, $\phi_{P2P\text{-}Q}$ and $\phi_{P2P\text{-}V}$, are similar to the values found for other regions (about 0.3 ln units). The remaining within-event single-path standard deviations, ϕ_{SP}, are quite small (about 0.2 ln units). For other regions, the ϕ_{SP} values are in the range of 0.23–0.30 [I–7]. The lower values ϕ_{SP} compared to previous studies may indicate some overfitting.

The correlation lengths for the non-ergodic terms are given in Table I–3. The long correlation lengths of 70 km for the site terms are due to the lack of site scaling in the reference ergodic GMM. This long correlation length reflects the spatial dimension of the average site condition. For the $\delta P2P_Q$ term, the correlation length of 80 km indicates the spatial variability of the crustal attenuation structure. For the $\delta P2P_V$ term, the correlation length was fixed at the value from the

French dataset of Ref. [I–8] because the compiled dataset is not sufficient to constrain this correlation.

TABLE I–3: CORRELATION LENGTHS FOR THE NON-ERGODIC TERMS IN THE GMM.

Term	Correlation length	
	ln (PGA)	ln (PGV)
$\delta S2S_{reg}$	73 km	73 km
$\delta P2P_Q$	80 km	80 km
$\delta P2P_V$	40 km*	40 km*

* Fixed value from French ground motion data

The PGA residuals for Source 1 (Fig. I–2) are shown in Fig. I–3. There are large positive residuals in the north direction for this source which is consistent with the attenuation anomaly seen in the intensity data (Fig. I–1). The non-ergodic GMM captures the median effect of this attenuation anomaly along with the epistemic uncertainty in the attenuation anomaly due to the limited data number of path paths shown in Fig. I–2. The non-ergodic GMM framework allows the attenuation anomaly to be included in seismic hazard studies for this region.

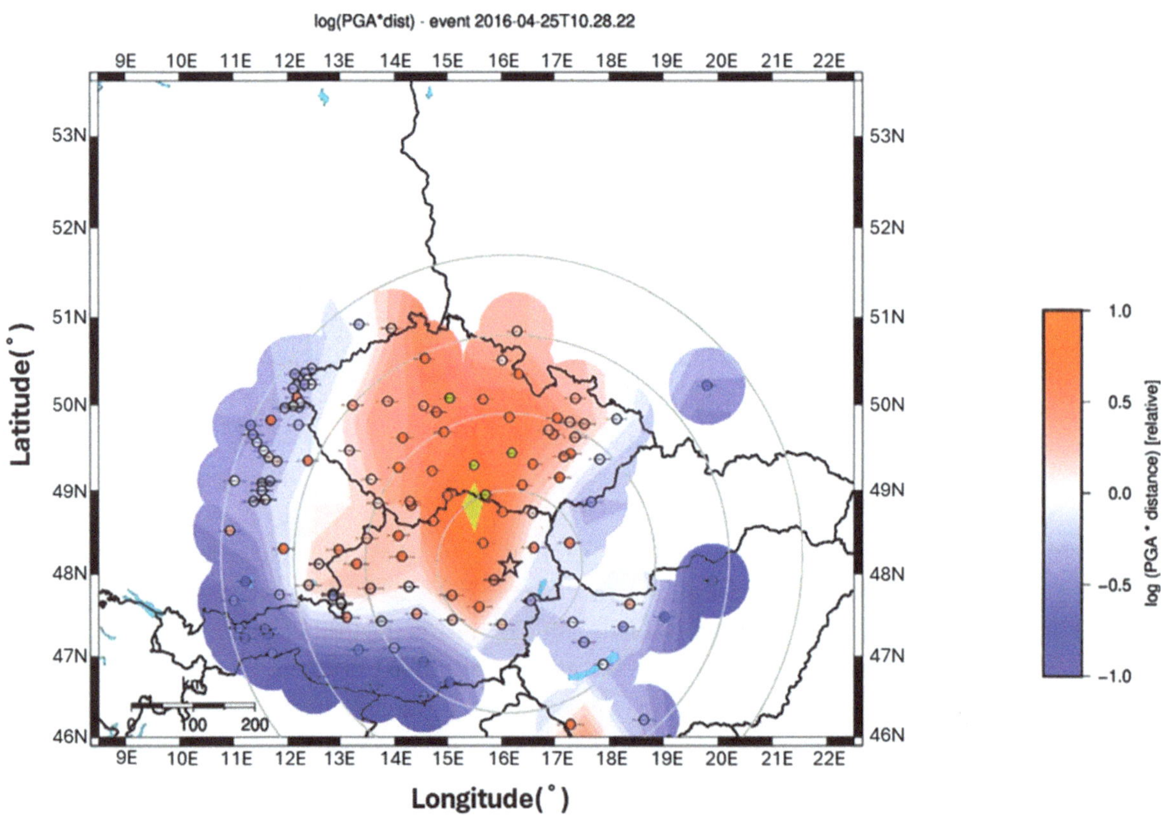

FIG. I–4. Spatial distribution of the PGA residuals for Source 1 shown in Figure I–2. Horizontal and vertical axes are the coordinates in longitude and latitude in degrees, respectively.

An alternative way to show the non-ergodic effects the distance scaling for a given site. The distance scaling of the reference ergodic and non-ergodic models is compared in Fig. I–5 for three selected sites: Site 1 and Site 2 are near the Dukovany and Temelín nuclear installation sites and Site 3 is close to the well recorded earthquakes that constrain the short-distance attenuation. The left-hand frames in Fig. I–5 show the median non-ergodic terms for each site for PGA and PGV. The median attenuation for the ergodic GMM, partially non-ergodic GMM (site only), and non-ergodic GMMs are shown in the right-hand frame in Fig. I–5. Figure I–5 shows that including only the non-ergodic site term leads to a constant shift in the distance scaling as compared to the ergodic GMM. Including the non-ergodic path terms, particularly the $\delta P2P_Q$ affects the attenuation at the large distances (> 50 km), but it is a relatively small effect. The $\delta P2P_V$ term has a much larger effect and that can be seen at both short and long distances.

The epistemic uncertainty in the non-ergodic terms needs to be included in PSHA applications that implement non-ergodic GMMs. The epistemic uncertainty is computed by generating multiple sets of the spatially correlated non-ergodic terms that sample the epistemic uncertainty from the variable coefficient model for non-ergodic terms (see Section 8 for details). The epistemic uncertainty in the distance attenuation for 100 realizations of the non-ergodic terms is presented in Fig. I–6 by pink lines. The size of the epistemic uncertainty ranges from a factor of 2–4 in PGA. This range is related to the density of the data along a particular path. For PSHA applications, this epistemic uncertainty will be included in the logic tree for the non-ergodic GMM.

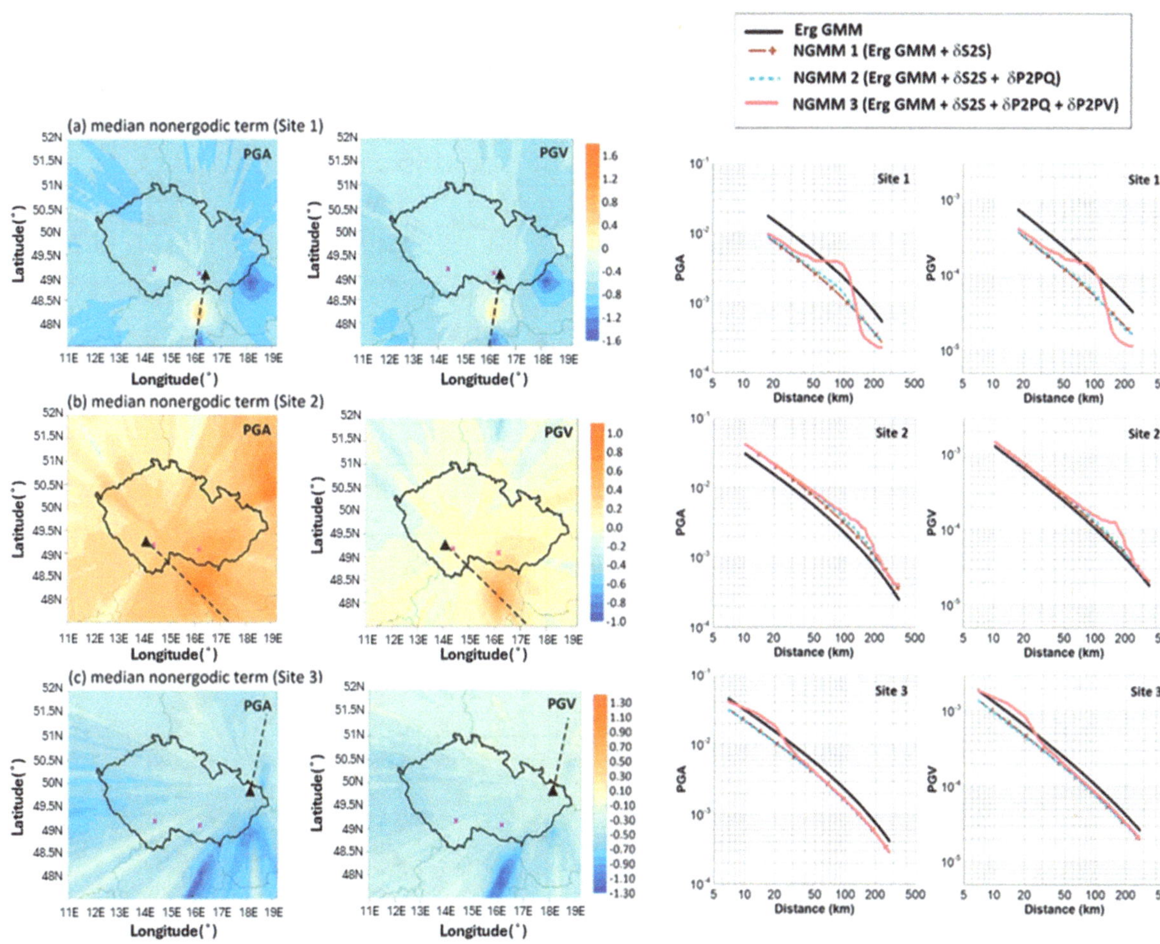

FIG. I–5. Example of the median non-ergodic terms and the median attenuation with distance in the indicated direction for three sites. Horizontal and vertical axes are the coordinates in longitude and latitude in degrees, respectively.

I-3. CONCLUSION

The evaluation of the attenuation anomalies in the Bohemian Massif clearly showed that the application of site specific attenuation relations based on non-ergodic ground motion model is a suitable way to reduce epistemic uncertainties in PSHA. The PSHA step of selecting attenuation relations needs to include mapping the issue of attenuation anomalies using all available data, both from the pre-instrumental period and precise instrumental data and deciding how significantly this phenomenon contributes to the seismic load of the evaluated area or certain nuclear power plant. Local seismic networks monitoring the vicinity of nuclear power plants and national seismic networks in the region of interest are in this case an irreplaceable source of quality data.

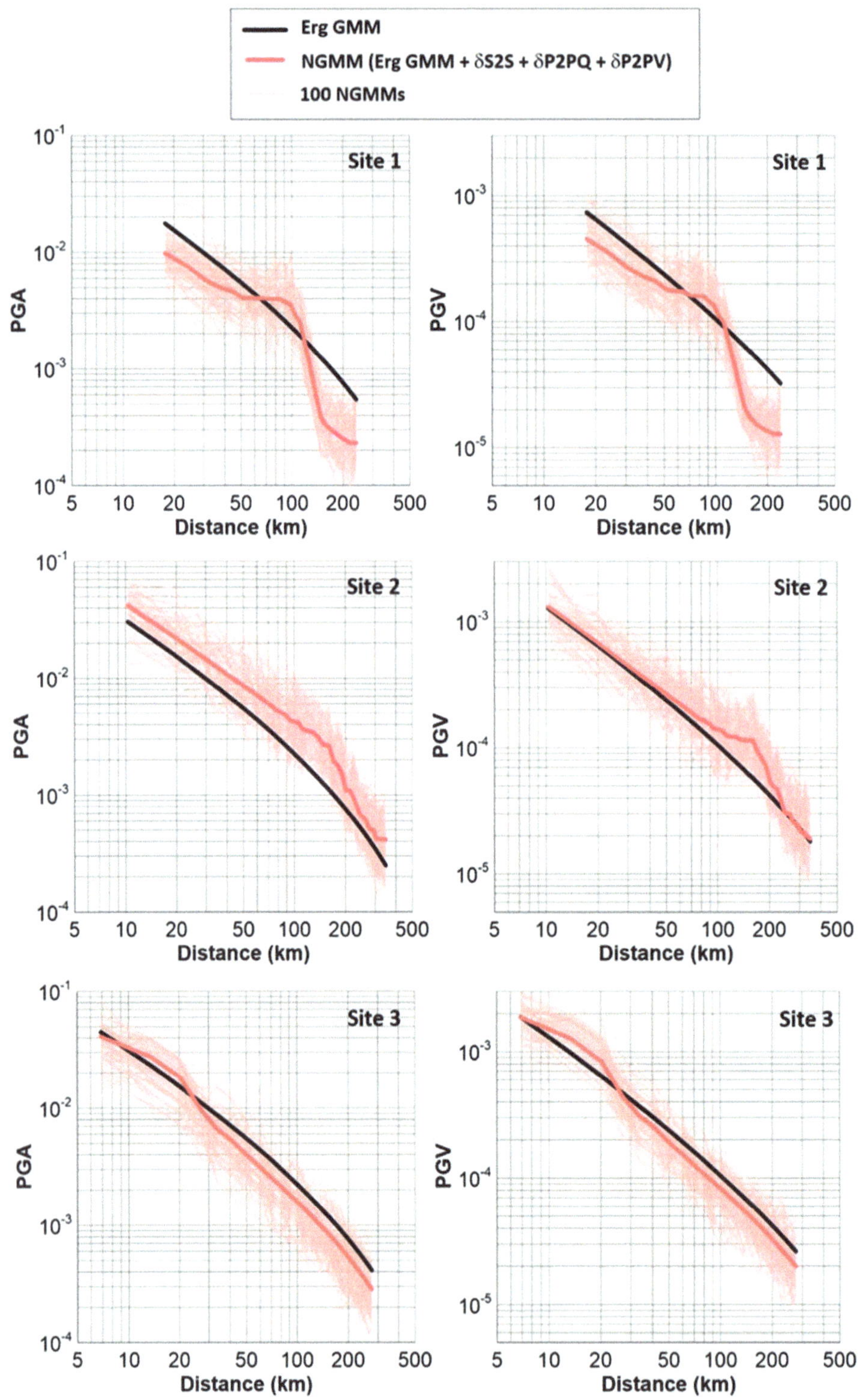

FIG I–6. Example of the epistemic uncertainty in the attenuation with distance for the three sites and directions shown in Figure I–4.

REFERENCES TO ANNEX I

[I–1] ZÁTOPEK, A., The propagation of East Alpine earthquakes in the Bohemian Mass, Publications de l'Institut Géophysique National Tchécoslovaque Travaux Spéciaux **3** (1948) 48.

[I–2] SCHMIDT, J. J. F., Charte der Verbreitung und Intensität des Erdbebens am 15. Jänner 1858 gezeichnet von J.F. Jul. Schmidt. Taf. II. Color. lithograph, **23**,5 x 28 cm, k. k. geographische Gesellschaft, Wien, 1858.

[I–3] PROCHÁZKOVÁ, D., KÁRNÍK, V., (Editors). Atlas of Isoseismal Maps for Central and Eastern Europe. Geophysical Institute of the Czechoslovak Academy of Sciences. Prague, 1978.

[I–4] MÁLEK, J., BROKEŠOVÁ, J., VACKÁŘ, J., Mid-European seismic attenuation anomaly, Tectonophysics 712-713 (2017) 557–577, https://doi.org/10.1016/j.tecto.2017.06.003.

[I–5] MÁLEK, J., VACKÁŘ, J., PRACHAŘ, I., ŠPAČEK, P., PSHA of the ETE / EDU Site, Technical Report, Institute of Rock Structure and Mechanics of the Czech Academy of Sciences, Prague (2022).

[I–6] GEOPENTECH, Southwestern United States Ground-Motion Characterization, SSHAC Level 3 - Technical report (Rev 2) (2015).

[I–7] LIN, P-S., et al., Repeatable source, site, and path effects on the standard deviation for empirical ground-motion prediction models, Bulletin of the Seismological Society of America 101(5) (2011) 2281-2295, https://doi.org/10.1785/0120090312.

[I–8] SUNG, C-H. et al., A non-ergodic ground-motion model of Fourier amplitude spectra for France, Bulletin of Earthquake Engineering 21(11) (2023) 5293-5317, https://doi.org/10.1007/s10518-022-01403-1.

ANNEX II. NEW REGULATIONS FOR CONSIDERATION OF NON-ERGODICITY OF GROUND MOTIONS AND THE STATE-OF-THE-PRACTICE IN JAPAN

Deterministic seismic hazard assessment approach has been applied to develop the design basis ground motions of nuclear installations in Japan. Nevertheless, consideration of the non-ergodicity of ground motions is required by the Nuclear Regulatory Authority of Japan (NRA) regulations and recommended by scientific communities, therefore some of the methods defined in this publication are included in the current practice of the Japanese utilities. This annex is composed of three parts: describing the NRA regulations regarding non-ergodicity in Section II–1; introducing the practice applied by the Japanese utilities in Section II–2; and summarizing a community- and government-recommended methodology for National Seismic Hazard Maps of Japan in Section II–3.

II-1. NRA'S REGULATORY STANDARDS ON DESIGN BASIS GROUND MOTIONS

The NRA regulations [II–1] require that the design basis ground motions are formulated as horizontal and vertical component ground motions at the free surface of a competent layer. Two distinct types of ground motion are defined as; (i) ground motions to be formulated by identifying seismic sources for each site and (ii) ground motions to be evaluated without identifying seismic sources. For the former type of ground motion, multiple earthquake scenarios are selected by considering the differences in the earthquake occurrence pattern (i.e. continental crust, inter-plate, and oceanic intra-plate earthquakes). Both the fault rupture modelling and the response spectrum approaches (i.e. GMPE) are applied to selected earthquake scenarios. In response spectra approach, the characteristics of the site (with various uncertainties) are taken into consideration and the response spectrum is formulated on the basis of near source records from continental crust earthquakes that are difficult to relate to known capable faults.

Though ergodicity and non-ergodicity of ground motions have not been explicitly mentioned, NRA regulatory standards specify treatments of the factors (e.g. source and site effects) that are supposed to affect the ground motion evaluation significantly. In particular, the NRA regulations require that the ground motions to be formulated by identifying seismic sources for each site are estimated by taking the regional characteristics into consideration, such as the earthquake occurrence pattern and seismic wave propagation, on the basis of ground motion records observed at the site. Regarding the characteristics of seismic wave propagation, the NRA regulations explicitly requires that the effects of 3-dimensional velocity structure around and beneath the site needs to be appropriately evaluated. NRA's relevant review guide [II–2] stipulates that the empirical Green's function method is applied as the fault rupture modelling approach, where seismic records are available and can be used as appropriate Green's functions from the viewpoint of record quality and fault source characteristics. It is expected that the application of the empirical Green's function method account for source, path, and site specific effects on ground motions. NRA's relevant review guide further stipulates that ground motions are appropriately evaluated by considering the characteristics of the applied GMM as well as the wave propagation effects (site effects) based on the subsurface structure model around the site. Moreover, the review guide states that seismic ground motion records, if available, are collected, analysed, and reflected in the estimates of response spectra by evaluating the effects due to earthquake occurrence pattern and regional/local characteristics of seismic wave propagation.

In addition, regarding the requirement of referring to the corresponding exceedance probability for both type of ground motions mentioned above, NRA's review guide stipulates that aleatory variability and epistemic uncertainty are included in ground motion evaluation.

II-2. UTILITIES' STATE OF THE PRACTICE IN JAPAN

This section presents the specific methods that are used in Japan for modelling the underground velocity structure. In addition, application examples for nuclear power plant (NPP) sites in Japan are presented. As explained in the previous section, the NRA regulations mandate the consideration of non-ergodicity in GMMs. This entails assessing the potential impacts arising from various sources (like the distinction between crustal earthquakes and subduction zone earthquakes), propagation paths (which may vary depending on region or site), and site specific amplifications. These assessments need to be conducted through detailed surveys and calibrated based on ground motion observation data, where possible.

Modelling of the underground velocity structure is one of key issues in evaluating the ground motions at the target site, which also affects the results of site response simulations. Therefore, modelling of the underground velocity structure is indispensable to constrain non-ergodic site terms using simulation-based methods presented in Section 7.

II-2.1. Underground velocity structure evaluation techniques

This section outlines the surveys using geophysical exploration methods and seismic records in order to model the underground velocity structure as recommended by the NRA's review guide [II–2]. Geophysical exploration methods include:

- Deep borehole surveys;
- Seismic exploration;
- Vertical seismic profile (VSP) extrapolation;
- Microtremor exploration.

Deep borehole survey methods use two main techniques: the downhole method and intra-borehole excitation and receiving method. In the downhole method, P-waves and S-waves generated by an artificial seismic source on the ground surface are collected by sensors in the borehole. P-wave and S-wave velocities are calculated from the travel distance and travel time of the waves between the source and the receiver. Therefore, this method may also be classified as a type of VSP exploration that, as described later, acquires elastic waves from an artificial seismic source on the ground surface, inside a borehole. In the intra-borehole excitation and receiving method, both the oscillator and receiver are lowered down into a borehole. In other words, P-wave and S-wave velocities at each depth are measured by lowering a probe that combines an oscillator and a receiver down a borehole. Techniques that are mainly used include the suspension method and dipole shear sonic imager.

The downhole method can estimate the velocities for the layers in the profile that may be defined using geological cross-sections. On the other hand, the intra-borehole excitation and receiving method can obtain better-constrained seismic velocities at each depth. By using a borehole, it is also possible to investigate the density of the layers, that are required along with the elastic wave velocities in the evaluation of the ground motion. This is called the density logging, where the density of the rock is estimated from the intensity of the gamma rays emitted from the radiation source in the logging equipment, that propagates through the geological strata and reaches the detector.

Seismic exploration methods assess the underground velocity structure through the observation of elastic waves produced by artificial seismic sources. These methods utilize multiple receivers deployed in two or three dimensions. Seismic exploration methods are typically categorized into seismic reflection and refraction exploration techniques. While both techniques employ artificial seismic sources, reflection exploration primarily focuses on determining the configuration of underground discontinuities, whereas refraction surveying is utilized for evaluating the elastic wave velocities. In certain cases, particularly in marine environments, high resolution 3-D seismic reflection exploration, originally developed for resource exploration, may be employed during reflection surveys. One of the requirements is that the site needs to facilitate the plotting of 2-D survey lines, or in other words, facilitate deployment of enough number of survey lines. Therefore, at a site where multiple buildings and structures already exist, it may not be possible to achieve the target depth or perform the exploration at a target location.

VSP is a technique for estimating the velocity structure by observing elastic waves generated by artificial seismic sources on the ground surface from inside a borehole. The methods of exploration include zero-offset VSP, offset VSP and walk-away VSP. In the zero-offset VSP, the seismic source is located near the borehole (often included in the family of downhole methods). In the offset VSP, the seismic source is located away from the borehole. In the walk-away VSP, observations are made while moving the seismic source in the horizontal direction. VSP combines the advantages of seismic exploration and deep borehole surveys. In contrast to P-S logging that uses only the initial motion, which also requires a borehole, the VSP uses reflected and transmitted waves. In addition, the 2-D velocity structure is eventually obtained in the offset VSP and walk-away VSP. However, for this method, the site needs to facilitate the deployment of long enough survey lines in 2-D. Therefore, if some buildings and structures already exist on a site, then there could be restrictions to the target depth and the possibility of performing the survey.

The earth experiences constant vibrations from various sources like traffic and ocean waves, resulting in tiny tremors known as microtremors. These microtremors consist of predominantly surface waves, whose characteristics reflect the underground velocity structure at their observation points. One common exploration technique that utilizes microtremors is the microtremor array method. This method estimates the phase velocities of surface waves within a designated array area by analysing microtremor records collected simultaneously from multiple locations. Subsequently, it deduces one-dimensional velocity structures through inversion analysis of these phase velocities [II–3]. Another approach involves the calculation of the ratio of horizontal to vertical ground motion spectrum, referred to as the H/V spectrum, using the microtremor records from a single observation point. The underground velocity structure is then inferred through inversion analysis, utilizing the ellipticity of Rayleigh waves [II–4].

Methods based on seismic records include:

- — Methods using seismic records at a single point on the ground surface;
- — Methods using vertical array seismic observations;
- — Methods using horizontal array seismic observations;
- — Seismic interferometry.

Commonly preferred methods for evaluating the underground velocity structure utilize the seismic records at a single point on the ground surface by using the receiver function [II–5], the H/V spectrum in the P-wave and coda wave [II–6], or diffuse wave field theory of the H/V spectrum [II–7]. These methods extract the respective quantities related to the underground velocity structure from seismic records and estimate 1-D velocity structure by inversion analysis. The receiver function is obtained by deconvolution of the radial component of the P-wave portion of the seismic records from the vertical component. The H/V spectrum obtained from seismic records is the spectral ratio of horizontal to vertical motion, as in microtremor analysis.

The analysis using seismic records from vertical arrays estimates the velocity and attenuation structures by obtaining the transfer functions from the stations in the array and fitting them with the theoretical transfer functions obtained by inversion analysis based on 1-D wave theory [II–8]. In many cases of inversion analysis, a theoretical amplification factor assuming normal incidence of P- or S-waves is used; while in some cases, the oblique incidence of P- or S-waves may be considered. Since attenuation parameters are less likely to be investigated in other geophysical explorations, damping ratios identified by this technique are often adopted for site response evaluation based on the transfer function approach [II–9].

The seismic records in a horizontal array may be used to examine the characteristics of the seismic waves propagating within the study area by focusing on long-period components in the observation records to understand the subsurface velocity structures [II–10]. In general, for surface waves which are the main component of long-period seismic ground motion, the phase velocities for each period are obtained by applying the frequency-wavenumber spectrum analysis [II–11] or semblance analysis [II–12]. Obtained phase velocities are either inverted in the same way as microtremor array exploration or validated by comparing them with the theoretical phase velocities from a proposed underground velocity structure model.

Seismic interferometry is a technique of extracting the Green's function from the cross-correlation functions of seismic records obtained at two observation points, which can synthesize a seismic record with the virtual source-receiver pair. A pseudo long-period earthquake ground motion record can be generated by stacking the cross-correlation functions between two points from microtremor recordings [II–13]. By analysing the dispersion of surface waves using the generated pseudo long-period earthquake ground records, it is possible to estimate the group velocity (reflecting the average underground velocity structure) between the two observation points. If seismic interferometry is applied to microtremor records obtained from a common reference point and array observation points, it is possible to synthesize pseudo long-period motion array records with the common reference point as the virtual source. Estimating the phase velocity based on long-period earthquake ground motion is also possible in areas where seismic records cannot be obtained [II–14, II–15].

II-2.2. Underground velocity structure modelling at nuclear power plants in Japan

To meet the regulatory requirements, utilities of all NPPs) in Japan have been investigating the underground velocity structures using various methods described in the previous section. The utilities have also been providing explanations to the regulatory during the conformity assessments. In this section, details of the surveys and analysis performed for evaluating the underground velocity structure for the Kashiwazaki-Kariwa NPP site are summarized.

At the Kashiwazaki-Kariwa NPP, seismic activities originating from the F-B fault in the sea and the western boundary fault zone of the Nagaoka Plain in the land were used as the focal earthquake scenarios. Ground motion simulations for the focal earthquake scenarios utilized the Empirical Green’s Function method. Therefore, modelling the underground velocity structure was not mandatory for the evaluation of the design basis ground motion. However, ground motions recorded at the NPP site during the 2007 Niigata Chuetsu-oki earthquake (a magnitude 6.8 earthquake with 17 km depth), were significantly higher than the design spectra of the nuclear installation (as of 2007). In addition, the maximum acceleration values recorded at the Arahama area (Units 1–4) were twice as high as the maximum acceleration values recorded at the Ominato area (Units 5–7). On-site vertical arrays have recorded ground motions since 1980s from 250 m below Tokyo Bay mean sea level (TMSL) up to the ground near Unit 1 and from 300 m below TMSL up to the ground near Unit 5. Following the 2007 Niigata-ken Chuetsu-oki Earthquake, various underground velocity structure surveys have been conducted to evaluate the cause of the large accelerations and spatial variation of accelerations at the site. A dense horizontal array of 20 seismometers (later increased to 30 seismometers) has also been deployed at the site. Table II–1 lists the surveys used in modelling the subsurface structure at the Kashiwazaki-Kariwa NPP.

At the northern part of the site, the geologic deposits are characterized by the Miocene to Lower Pliocene Shiiya Formation and is surrounded by the lower Pliocene to Pleistocene Nishiyama Formation. The geologic deposits at the southwestern part of the site consists of the Haizume Formation which is surrounded by the lower Nishiyama Formation. The geologic structure is characterized by a fold, found in the strata below the Nishiyama Formation, and it has been confirmed that the Madonozaka syncline is distributed between the Ushirodani anticline in the NE-SW direction and Nagamine anticline. Therefore, the hypothetical free surface basement depth changes from the north (134 m below TMSL near Unit 5) to the south (284 m below TMSL near Unit 1) of the site. P-S logging results properly reflect the site amplification factors below the hypothetical free surface basement level, as the geological layers with an S-wave velocity of 0.7 km/s or higher were distributed almost horizontally beneath that level. Deconvolution analyses using shallow subsurface model from the transfer function inversion confirmed that the seismic motions deconvolved to the hypothetical free surface basement on the Arahama area in the south and the Ominato area in the north for the 2007 Niigata-ken Chuetsu-oki earthquake were very similar.

The old fold structures were found in and around the Kashiwazaki-Kariwa NPP and the stratum boundaries of the fold structures were interpreted based on the results of seismic reflection surveys. In areas where the stratum boundaries cannot be identified by the surveys, the geologic horizons were estimated using the balanced cross-section method. In constructing the 2-D subsurface models, several 2-D subsurface structure surveys were overviewed in three dimensions as shown in Fig. II–1. Constructed 2-D models were validated by 1-D models that were built using joint inversion results based on H/V spectra of P- and S-coda wave parts and the receiver functions from the horizontal array seismograms. These were also compared with S-wave velocity structures from microtremor array surveys and P-S logging results at deep boreholes (Fig. II–2).

As shown in Fig. II–3, the site amplification factors estimated by finite element analysis using the 2-D underground velocity structure models show the differences in the incident wave with azimuth from the third asperity of the Niigata Chuetsu-Oki earthquake at Unit 1 (Arahama area) and Unit 5 (Ominato area). Unit 1 has a significantly larger site amplification. On the other hand, there was no significant difference in amplification characteristics between Unit 1 (Arahama area) and Unit 5 (Ominato area) from arrival directions other than the third asperity, such as the azimuth from the second asperity as shown in Fig. II–3 (lower right panel).

TABLE I–1. INVESTIGATIONS FOR THE UNDERGROUND VELOCITY STRUCTURE MODELLING AT THE KASHIWAZAKI-KARIWA NPP SITE.

Type of Investigation	Main use	Matter of Evaluation
Geological structure evaluation	Verification of fold structures, evaluation of 2-D structures, and identification of the strata boundaries of fold structures by the balanced cross-section method.	— Hypothetical free surface basement — Horizontally stratification — 2D and 3D structure
P-S logging	Estimating the depth of hypothetical free surface of the basement and the initial S-wave velocity model for transfer function inversion	— Hypothetical free surface basement — Shallow subsurface model
P-S logging (deep borehole)	Evaluation of the velocity structures down to the depth of 1 km.	— Shallow subsurface model — Deep subsurface model — Model validation
Seismic reflection survey	Estimation of 2-D structures and verification of fold structures.	— 2D and 3D structure
Microtremor array exploration	Estimation of S-wave velocity structures down to a depth of about 4 km.	— Model validation
Receiver function analysis and H/V spectrum	Estimation of subsurface model down to the bedrock as deep as about 4 km (joint inversion)	— Shallow subsurface model — Deep subsurface model — 2D and 3D structure
Transfer function	Evaluation of subsurface model down to around 300 m below TMSL.	— Shallow subsurface model
Check for the azimuthal dependency of observed ground motions and comparison between the units	Evaluation and comparison of amplification factors using vertical and horizontal arrays and the recordings of the Niigata-ken Chuetsu-oki earthquake.	— Local amplification
Site amplification factors	Evaluation and comparison of inland earthquakes and offshore earthquakes	— Local amplification
2D and 3D analysis	Evaluation of azimuthal dependency of spatial variation of amplification factors	— Model validation

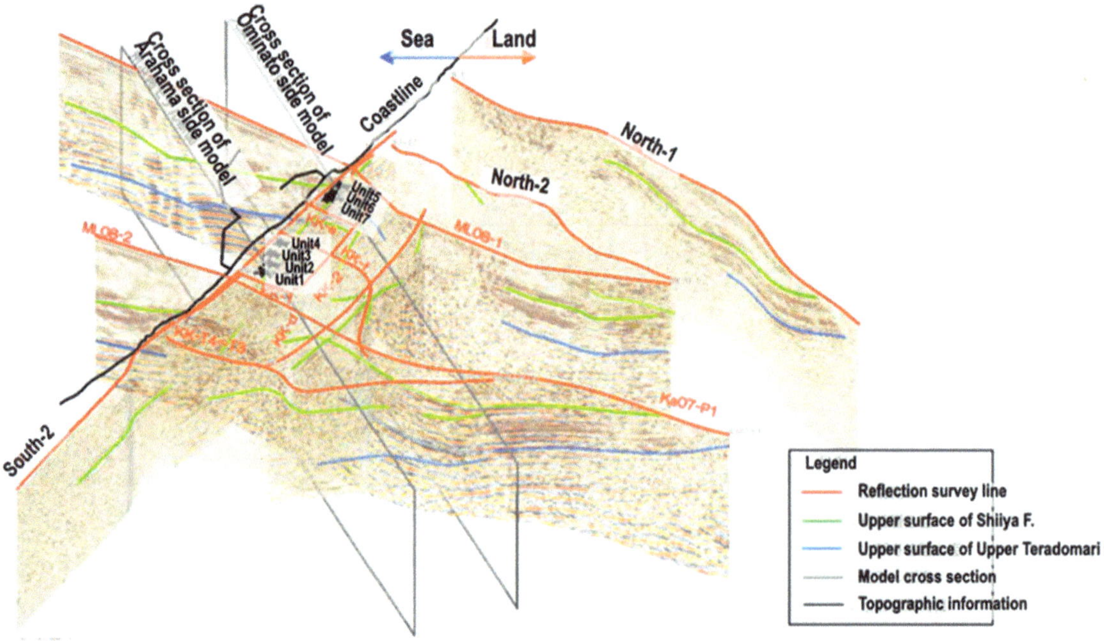

FIG. II–1. Results of seismic reflection surveys and 2-D underground velocity structure models of the Kashiwazaki-Kariwa NPP (reproduced from [II–16] with English notation added).

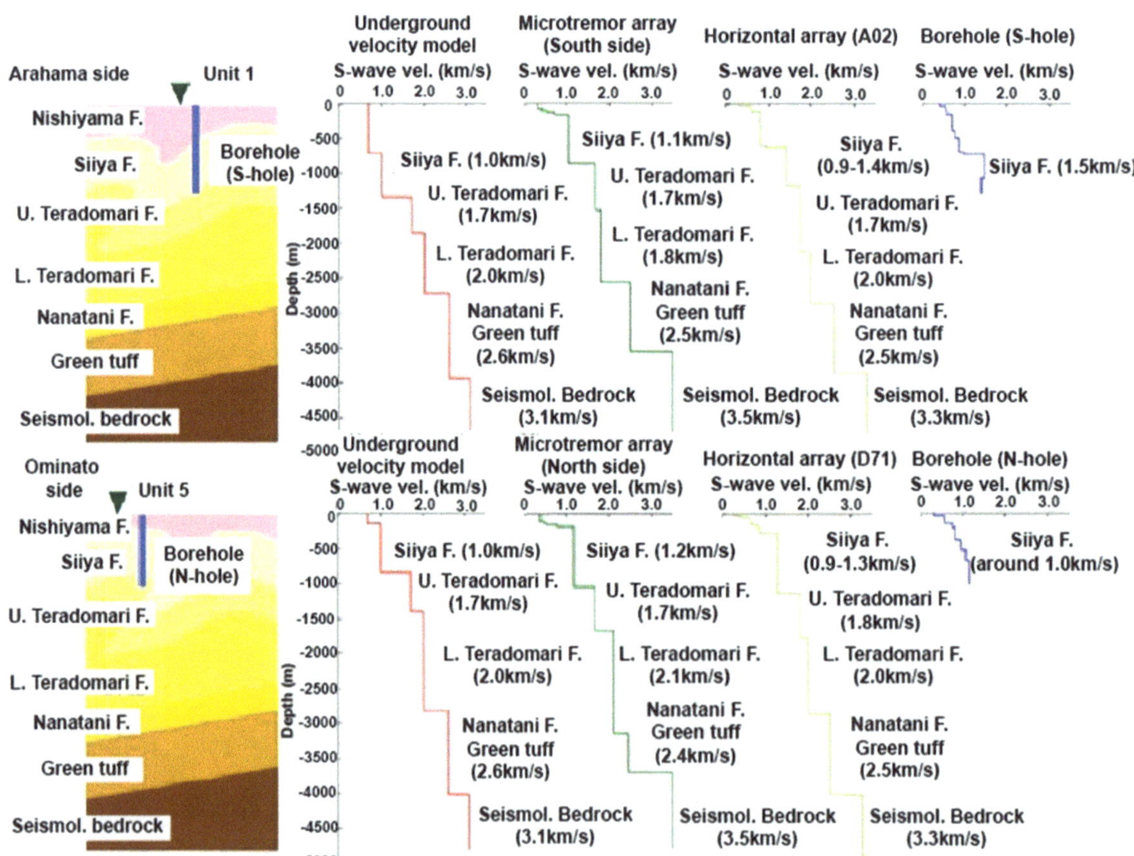

FIG. II–2. Comparison of the 2-D underground velocity structure model of the Kashiwazaki-Kariwa NPP and the 1-D underground velocity structure based on the various survey methods (reproduced from [II–16][42] with English notation added).

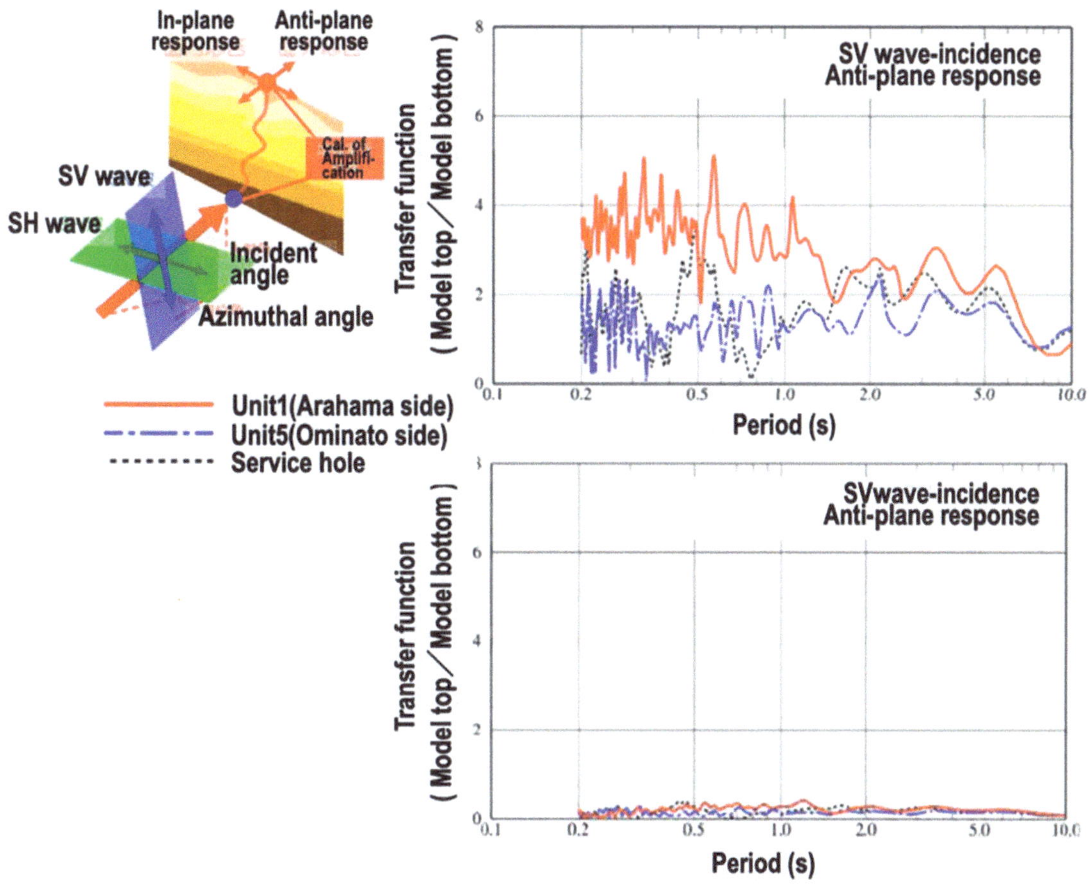

FIG. II–3. Comparison of amplification characteristics of 2-D model of the Kashiwazaki-Kariwa NPP base incidence from different azimuth angles (upper right: Incidence from azimuth of the third asperity, lower right: Incidence from azimuth of the second asperity (reproduced from [II–16] with English notation added).

To comply with the conformity assessments for the new regulatory requirements in Japan, various surveys have been conducted at NPPs to construct underground velocity structure models that are used to formulate the design basis earthquake ground motion. As of the end of 2019, many NPPs have completed the conformity assessments related to the evaluation of the underground velocity structures for formulating the design basis ground motions. Overview of the conformity examinations indicates that the survey and model construction methods for underground velocity structures vary, depending on the site specific conditions of the NPP site such as geology, topography, seismicity, and the size of a site area. The surveys conducted at each NPP site under these conditions and physical constraints are summarized in Table II–2 and Table II–3.

TABLE II–2. LIST OF SURVEYS PERFORMED FOR UNDERGROUND VELOCITY STRUCTURE MODELLING AT DIFFERENT NUCLEAR POWER PLANTS IN JAPAN (ADAPTED FROM [II–16]).

Type of survey	Nuclear Power Plant Site						
	Tomari	**Onagawa**	**Tokai 2**	**Kashiwazaki-Kariwa**	**Hamaoka**	**Mihama**	**Ooi**
Geological structure evaluation	✓	✓	✓	✓	✓	✓	✓
Seismic prospecting at pilot tunnels						✓	✓
P-S logging	✓	✓	✓	✓	✓	✓	✓
P-S logging (deep borehole)			✓	✓	✓	✓	
Seismic reflection / refraction survey	✓		✓	✓	✓	✓	✓
Offset vertical seismic profiling (deep borehole)					✓		
Microtremor array extrapolation			✓	✓	✓	✓	✓
Seismic interferometry							✓
Dense microtremor observation			✓			✓	✓
Earthquake record analysis	✓	✓	✓	✓	✓	✓	
2D-3D amplification evaluation	✓	✓	✓	✓	✓		

TABLE II–3. LIST OF SURVEYS PERFORMED FOR UNDERGROUND VELOCITY STRUCTURE MODELLING AT DIFFERENT NUCLEAR POWER PLANTS IN JAPAN (cont.'d) (ADAPTED FROM [II–16]).

Type of survey	Nuclear Power Plant Site						
	Takahama	**Shimane**	**Ikata**	**Genkai**	**Sendai**	**Ohma**	**Tsuruga**
Geological structure evaluation	✓	✓	✓	✓	✓	✓	✓
Seismic prospecting at pilot tunnels	✓		✓	✓	✓		✓
P-S logging	✓	✓	✓	✓	✓	✓	✓
P-S logging (deep borehole)		✓	✓			✓	✓
Seismic reflection / refraction survey	✓	✓	✓		✓	✓	✓
Offset vertical seismic profiling (deep borehole)		✓	✓			✓	✓
Microtremor array extrapolation	✓	✓		✓	✓	✓	✓
Seismic interferometry	✓					✓	✓
Dense microtremor observation	✓	✓		✓	✓	✓	✓
Earthquake record analysis		✓	✓	✓	✓	✓	✓
2D-3D amplification evaluation		✓	✓	✓	✓	✓	

II-3. PROCEDURES TO BUILD A SUBSURFACE VELOCITY STRUCTURE MODEL

This provides a summary of Ref. [II–17] (published recently by the Headquarters for Earthquake Research Promotion (HERP) in Japan) that describes the procedures for constructing the subsurface velocity structure model for the ground motion simulations used in the National Seismic Hazard Maps of Japan [II–18]. The subsurface velocity structure model defined by HERP is a three-dimensional (3-D) velocity structure that includes main physical parameters such as density, P-wave and S-wave velocities, crustal attenuation values, and the boundaries for each layer. The model is divided into three domains: layers of the crustal structure, deep sedimentary layers, and shallow soil layers as shown in Fig. II–4. The crustal structure influences the characteristics of the seismic wave propagation from the earthquake sources and includes layers down to the upper mantle. Deep sedimentary layers influence the amplification of seismic waves in a wide period range (0.1–10 seconds) that are of interest in the evaluation of broadband ground motions. Shallow soil layers influence the amplification of seismic waves in a short-period range (less than 2 seconds). The methods and data for modelling each domain are provided following sub-sections.

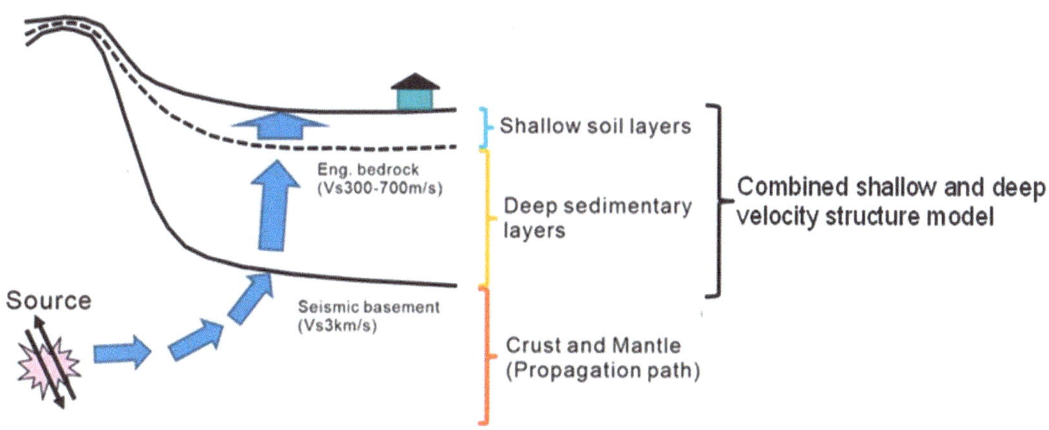

FIG. II–4. Schematic diagram of the domains of velocity structure model (taken from [II–17] with permission)

II-3.1. Layers of crustal structure

Seismic velocity models beneath the top surface of the seismic bedrock, including the layers in the upper mantle, are typically constructed by using the P-wave and S-wave travel times of earthquake records, the results of deep borehole surveys and the physical properties obtained with seismic reflection and refraction surveys. To estimate the spatial distribution of the top surface of the seismic basement that has an S-wave velocity higher than 3 km/s, gravity anomaly maps are used together with the relationship between the basement depth and gravity anomaly values. The Conrad discontinuity and the Mohorovičić discontinuity are modelled using the results from the seismic surveys and the regional gravity anomaly distributions. Hypocentre data, finite fault models, and seismic tomography models are also used. The 3-D velocity structure model beneath the seismic basement is constructed by interpolating the depths of the seismic basement and the velocity structure obtained from the original data together with the S-wave velocities of the referred models. The use of a seismic tomography model can be effective for introducing complex velocity information, such as mantle wedges and subducting slabs, into the crustal and mantle structure model.

II-3.2. Deep sedimentary layers

Information including regional geological maps, geological cross sections, soil profiles obtained by deep boreholes (e.g. borehole observatory for crustal activity), results from various geophysical surveys (seismic reflection and refraction surveys, microtremor surveys, gravity surveys, and electromagnetic surveys) including P-S logging, and studies using earthquake ground motion records are compiled to develop the 3-D velocity model of deep sedimentary layers.

II-3.2.1. 1-D velocity structure models

Existing geological maps, stratigraphic sections, borehole data, and regional geological stratigraphic data are used to construct the 1-D soil profiles that are composed of multiple strata. The velocity structure for the 1-D profiles are estimated by comparing the P-wave velocities obtained in various surveys with the geological classifications. The relationships between the geological classifications and the physical properties that were established in previous studies may be useful (e.g. Ref. [II–19]). The S-wave velocities, densities, and crustal attenuation values for individual layers may be determined by using the results of different surveys (e.g. S-wave velocities from P-S loggings and microtremor surveys, densities from density loggings, etc.) and analysis of earthquake ground motion records. If only the P-wave velocities are available, the S-wave velocities, densities, and crustal attenuation values can be derived from empirical and theoretical equations from the P-wave velocities. The relation proposed by Ref. [II–20] is an example of the relations between the physical values, from which one can estimate the S-wave velocity and density when both the P-wave velocity and void ratio are known. Empirical equations to convert the P-wave velocity to density are also available (e.g. Refs [II–21, II–22]). When using these theoretical and empirical equations, it is important to confirm their applicability by referring to the original data used to derive these equations.

II-3.2.2. Development of 3-D velocity structure model using 1-D profiles

To estimate a 3-D velocity structure model using multiple 1-D velocity structure models, layers with similar geological classifications in the 1-D profiles are categorized into the same velocity layer. For determining the spatial extent of each velocity layer, spatial interpolation is performed. In this procedure, boundaries of the layers with different velocities detected by seismic reflection and refraction surveys, densities estimated from the gravity anomaly distribution filtered at wavelengths corresponding to the target depth of the structures, existing geological cross sections, geological contour maps, geological ages, and shapes of the faults and the folding structures are considered. Based on the 3-D distribution of individual geological boundaries, an initial model of the deep sedimentary layers from the top of the engineering bedrock to the top of the seismic basement is built. At the top of the engineering bedrock, a stratum is chosen to have an S-wave velocity used in the definition of the engineering bedrock. In an area where the top of the seismic basement is very shallow, weathering layers near the surface over the basement are also considered. The final shape of each boundary can be refined through a forward analysis and an inversion of the observed gravity anomaly data.

II-3.2.3. Validation of 3-D velocity structure model using earthquake ground motion records and ground motion simulations

S-wave velocities and the depths to boundaries of deep sedimentary layers may be confirmed with earthquake ground motion records, based on surface wave dispersion analysis, horizontal-to-vertical spectral ratios, and receiver functions. Theoretical values calculated from the 3-D model may be compared with the observed values to validate the 3-D model. If necessary, the model is adjusted to have a better fit with the observed values.

Simulations of earthquake ground motions using the estimated 3-D velocity structure model of the deep sedimentary layers may be performed to validate the accuracy of the velocity structure model. In order to develop the subsurface velocity model for the Japanese seismic hazard maps, simulated seismograms calculated by analytical methods based on the elastic wave theory, such as the finite difference method, were compared with the observed ones to confirm whether the simulations accurately reproduce various characteristics of the observed waveforms, such as amplitudes, predominant periods, arrival times of distinct phases, and time-variant characteristics. The subsurface velocity models may be adjusted to enhance the consistency of simulations with the ground motion characteristics of the observed recordings. In this procedure, the crustal attenuation value of each layer, which controls the attenuation features of the seismic waves, need to be set. To establish a high accuracy crustal attenuation model, information related with the regionality (i.e. geological classification) of the physical properties for a target area may be referred to.

II-3.3. Shallow soil layers

Amplification of ground motions at short spectral periods (mainly for periods less than 2 s) is significantly affected by S-wave velocity structures of the shallow soil layers. The method to model the shallow soil layers is selected by considering the amount of data and computing resources. In Ref. [II–18], two methods for modelling shallow soil layers, based on geomorphologic classification model and the borehole data, are presented.

Using the geomorphic classification model, the region of interest is discretized into an appropriately sized mesh and the representative topographical and geomorphic units in each mesh are re-categorized with the given criteria. For the re-categorized meshes, the site amplification characteristics may be empirically estimated with explanatory variables such as altitude, slope angle, and distance from mountains or hills formed in old geological ages. The V_{S30} (the time-averaged S-wave velocity in the top 30 m from the surface) and the amplification factor may be estimated using empirical relationships based on P-S logging data and each geomorphologic classification. This approach was used to evaluate the site amplification characteristics throughout Japan, including areas with insufficient soil data [II–23].

When the borehole data is used, the locations of available boreholes are marked for each mesh and assigned with various attributes, such as the availability of related P-S logging data, drilling depths, and the geomorphic classifications. Then, a representative soil profile is developed for each mesh. When a mesh contains multiple boreholes, the deepest borehole or borehole in connection with the P-S logging measurement may be preferred. For this representative soil profile, layers may be classified by considering the geologic stratigraphy, soil classification methods, and in-situ geotechnical test results by using relations available in the literature. P- and S-wave velocities obtained by P-S logging may be assigned to the layer classifications. When no P-S logging data are available in a mesh, a velocity value may be estimated for each layer using an empirical equation. Many empirical relations from various types of in-situ geotechnical survey, soil type, geological age, and various depths are available to estimate the S-wave velocities (e.g. [II–24–II–27]). For a mesh without any borehole or P-S logging data, the 1-D velocity structure model of the representative soil profile of one of the neighbouring meshes that has the same geomorphologic classification may be used.

II-3.4. Building combined models for shallow and deep layers

A combined model of shallow and deep layers is constructed for evaluating broadband ground motion by combining the models of deep sedimentary layers and shallow layers without creating an artificial discontinuity. The model for shallow soil layers may be directly overlaid on the model for the deep sedimentary layers when the difference of the S-wave velocities, or the impedance contrast, between the bottom of the shallow soil layers model and the top of the deep sedimentary layers model is small (Fig. II–5). When the impedance contrast is large, intermediate layers having interpolated S-wave velocities are added between these two models to prevent an artificial discontinuity. The thicknesses and velocities of the additional intermediate layers are set by referring to existing data, such as P-S logging data from deep borehole surveys. The artificial discontinuity of two models in the 3-D model may be avoided with the adjustment of the thicknesses of the additional layers, considering the regional variation of the depth to the deep sedimentary layers and the depth to the top of the seismic basement in a wide region including basins or mountain areas (Fig. II–5).

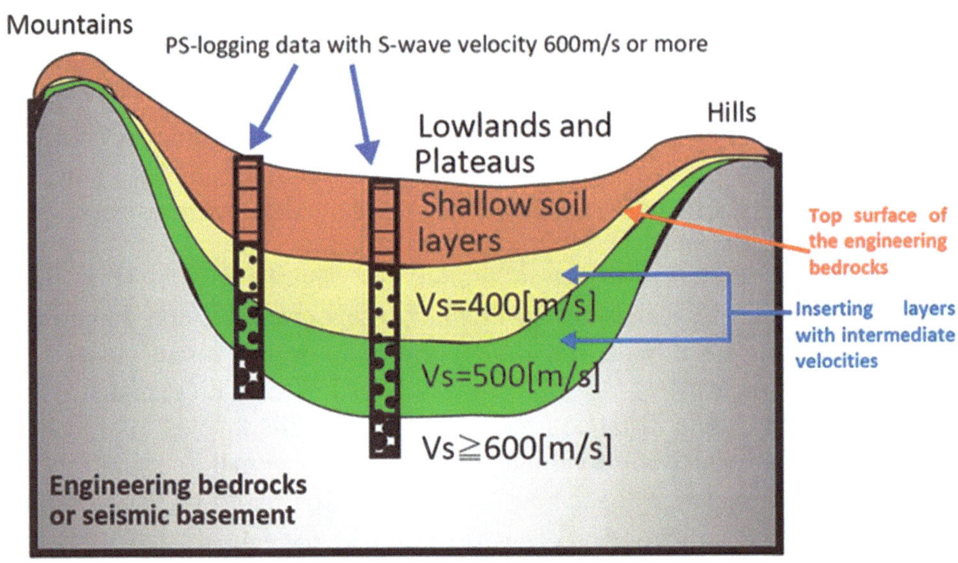

FIG. II–5. Schematic figure of additional layers with intermediate velocities defined to avoid discontinuities in the 3-D velocity model (taken from [II–17] with permission).

The combined model of shallow and deep layers is adjusted using information on the seismic velocity structures beneath the observation points that extracted from the earthquake records and microtremor data. If there is difficulty in fine-tuning the model, additional data may be collected or microtremor observations may be performed. Various analysis methods are available to extract information on seismic velocity structures from earthquake records and microtremor observation data, such as H/V spectrum, which is the Fourier spectrum ratio of the horizontal motion to vertical motion of earthquake records (called the R/V spectrum when applied to the Rayleigh wave part), H/V spectrum of microtremor data, spectral ratio of a seismic record on the surface to a borehole record in a vertical array, and surface wave phase velocity from a microtremor array survey. For example, the S-wave velocity and thickness of a layer can be estimated from the comparison of the peak period and shape of the observed H/V spectrum of microtremor data at a single point with the theoretical ellipticity of Rayleigh waves assuming a homogeneous and layered velocity structure.

After estimating the site amplifications at short periods (periods shorter than 2 s) by means of the empirical methods with ground motion records from moderate-to-small magnitude earthquakes, estimated amplifications may be used to validate the combined model of shallow and deep layers. The validated or adjusted combined 3-D model of shallow and deep layers may be used to simulate ground motions for moderate-to-small magnitude earthquakes to compare overall characteristics of the observed and simulated waveforms, such as P-S travel times, amplitudes, peak periods, spectral shapes, time-variant characteristics, and durations, may be compared. If there are inconsistencies, combined model of shallow and deep layers may be re-adjusted.

REFERENCES TO ANNEX II

[II–1] NUCLEAR REGULATION AUTHORITY OF JAPAN, Interpretation of the Regulations on Standards for the Location, Structure, and Equipment of Commercial Power Reactors and Auxiliary Facilities (2013, revised on February 22, 2023), https://www.nra.go.jp/data/000382455.pdf (in Japanese).

[II–2] NUCLEAR REGULATION AUTHORITY OF JAPAN, Review Guide on Design Basis Ground Motions and Seismic Design Policy (2013), https://www.nsr.go.jp/data/000069160.pdf (in Japanese).

[II–3] OKADA, H., MATSUSHIMA, T., MORIYA, T., SASATANI, T., Long-period microtremor exploration for wide-area and deep subsurface exploration, Geophysical Exploration **43** (1990) 402-417 (in Japanese with English abstract).

[II–4] ARAI, H., TOKIMATSU, K., S-wave velocity profiling by inversion of microtremor H/V spectrum, Bulletin of the Seismological Society of America **94** (2004) 53-63, https://doi.org/10.1785/0120030028.

[II–5] LANGSTON, C. A., Structure under Mount Rainier, Washington, inferred from teleseismic body waves, Journal of Geophysical Research: Solid Earth **84** (1979) 4749-4762, https://doi.org/10.1029/JB084iB09p04749.

[II–6] KOBAYASHI, K., et al., "Estimation of deep underground velocity structures by inversion of spectral ratio of horizontal to vertical component in p-wave part of earthquake ground motion" (Proc. of 12th World Conference on Earthquake Engineering), Auckland, New Zealand (2000) (in Japanese with English abstract).

[II–7] KAWASE, H., SANCHEZ-SESMA, F. J., MATSUSHIMA, S., The optimal use of horizontal-to-vertical spectral ratios of earthquake motions for velocity inversions based on diffuse-field theory for plane waves, Bulletin of the Seismological Society of America **101** (2011) 2001-2014, https://doi.org/10.1785/0120100263.

[II–8] OHTA, Y. "Application of Optimization Method to Earthquake Engineering 1. Estimation of Underground Structure of SMAC Observation Site in Hachinohe Harbor" (Proc. of the Architectural Institute of Japan) (1975) (in Japanese).

[II–9] SATO, H., Attenuation measurements of near-surface rock and its modelling for earthquake ground motion estimation, Geophysical Exploration **65** (2012) 37-51, (in Japanese with English abstract).

[II–10] YAMANAKA, H., ISHIDA, H., Application of genetic algorithms to an inversion of surface-wave dispersion data, Bulletin of the Seismological Society of America **86** (1996) 436-444, https://doi.org/10.1785/BSSA0860020436.

[II–11] CAPON, J., High-resolution frequency-wavenumber spectrum analysis, Proceedings of the IEEE **57** (1969) 1408-1418, 10.1109/PROC.1969.7278.

[II–12] NEIDELL, N. S., TANER, M. T., Semblance and other coherency measures for multichannel data, Geophysics **36** (1971) 482-497, https://doi.org/10.1190/1.1440186.

[II–13] WAPENAAR, K., FOKKEMA, J., Green's function representations for seismic interferometry, Geophysics **71** (2006) SI33-SI46, https://doi.org/10.1190/1.2213955.

[II–14] YAMANAKA, H., KATO, K., CHIMOTO, K., TSUNO, S., Estimation of surface-wave phase velocity from microtremor observation using an array with a reference station, Exploration Geophysics **46** (2015) 267-275, https://doi.org/10.1071/EG14069.

[II–15] SATO, H., et al., application of seismic interferometry to an exploration of subsurface structure by using microtremors in the Wakasa Bay Area, Report of Central Research Institute for Electric Power Industry, N14020 (2014) (in Japanese with English abstract), https://doi.org/10.5610/jaee.15.7_454.
[II–16] 266th Review Meeting on Nuclear Power Plant's Compliance with New Regulatory Standards: Comments and Responses on the Amplification Characteristics of Seismic Waves at the Kashiwazaki-Kariwa Nuclear Power Station's Units 6 and 7, (2015) (in Japanese).
[II–17] Earthquake Research Committee, the Headquarters for Earthquake Research Promotion, Procedures to Build a Subsurface Velocity Structure Model (2022) https://www.jishin.go.jp/main/chousa/17apr_chikakozo/model_concept-e.pdf
[II–18] Earthquake Research Committee, the Headquarters for Earthquake Research Promotion, National Seismic Hazard Maps for Japan (2020), https://www.jishin.go.jp/evaluation/seismic_hazard_map/shm_report/shm_report_2020/ (in Japanese)
[II–19] SUZUKI, H., Geology of the Koto Deep Borehole Observatory and Geological Structure Beneath the Metropolitan Area, Report of the National Research Institute for Earth Science and Disaster Prevention, Japan, (1996) (in Japanese).
[II–20] GASSMANN, F., Über die elastizität poröser medien, Vierteljahrss-chrift der Naturforschenden Gesellschaft in Zurich **96** (1951) 1-23.
[II–21] LUDWIG, W. J., NAFE, J. E., DRAKE, C. L., "Seismic refraction in the sea", Vol. 4, (Maxwell, A. E, Ed.), Wiley-Interscience, New York (1970).
[II–22] GARDNER, G. H. F., GARDNER, L. W., GREGORY, A.R., Formation velocity and density—The diagnostic basics for stratigraphic traps, Geophysics **39** (1974) 770-780, https://doi.org/10.1190/1.1440465.
[II–23] WAKAMATSU, K., MATSUOKA, M., Development of the 7.5-arc-second Japan engineering geomorphologic classification map based on uniform national standards and the use of the map, Bulletin of Japan Association for Earthquake Engineering **18** (2013) 33-38 (in Japanese).
[II–24] IMAI, T., TONOUCHI, K., Relationship between the N-value and the S-wave velocity and its usage examples, Journal of Foundation Engineering and Equipment **16** (1982) 70-76 (in Japanese).
[II–25] OHTA, Y. and GOTO, N., Estimation of S-wave velocity in terms of characteristic indices of soil, Butsuri-Tanko (Geophysical Exploration) **29** (1976) 251-261 (in Japanese).
[II–26] MASAKI, K., A study about evaluation of ground motion characteristics for earthquake disaster prevention, Doctoral dissertation at Tokyo Institute of Technology, 1984, 167 (in Japanese).
[II–27] FUKUWA, N., ARAKAWA, M., KOIDE, E., ISHIDA, E., Dynamic soil modelling based on existing soil data in urban area using GIS, AIJ Journal of Technology and Design **9** (1999) 249-254 (in Japanese).

ANNEX III. IMPLEMENTATION OF FULLY NON-ERGODIC GROUND MOTION MODELS FOR PSHA OF NPPs – SLOVENIA EXAMPLE

Developing and implementing PSHA at an NPP site is a complex, long term process. One major challenge is the need to minimize the impact of epistemic uncertainty on PSHA results. In 2015, the Slovenian Environment Agency identified that ground motion modelling significantly influences seismic hazard at the Krško NPP site. The GMMs developed since 2004 lead to increased ground motion intensity predictions, which was reflected in the increase of seismic hazard in terms of site specific PGA. However, these models are operated under the ergodic assumption, which means that for a given scenario, the median and aleatory variability of the ground motion were assumed to be the same for all sites within the same broad tectonic category. Such ergodic GMMs are not very precise in decomposing the effect of the uncertainty in the prediction of ground motion intensity into the part related to epistemic uncertainty and that related to the aleatory uncertainty (see Section 4 for details). This shortcoming may be particularly important when the ergodic GMMs are applied in the PSHA at a specific location and for long return periods, as for the case of NPPs. However, the growth of the number of ground motion recordings makes it possible to practically eliminate the ergodic assumption, develop a new non-ergodic ground motion model, and implement it into the PSHA as defined in this publication.

Based on the discussions summarized above, GEN energija (GEN) decided to support the development of a non-ergodic GMM for the Krško region in 2019 within the framework of an international project for the development of engineering solutions for the construction of new NPPs, led by the Faculty of Civil and Geodetic Engineering at the University of Ljubljana and carried out together with researchers from France and US. The non-ergodic GMM was developed by researchers from Faculty of Civil and Geodetic Engineering, University of Ljubljana and University of California, Berkeley. The first report on the non-ergodic model was submitted to GEN at the end of 2020 and was a subject of review by an international Peer Review Team Panel. The review process started in January 2021 and concluded in April 2021, issuing the Revision 1 report on the model. As part of the international peer review of the PSHA for Krško site, Revision 2 of the report on the non-ergodic GMM for Krško was issued on March 14, 2024 [III–1].

The non-ergodic GMM developed for the Krško region is briefly summarized in this Annex and used in a simplified PSHA that utilizes a single area source for illustrative purposes. This simplified PSHA aims to present the impact of the non-ergodic GMM on the hazard estimation for the site in comparison to the selected ergodic GMM. Consequently, the presented results in this Annex do not reflect the hazard at the Krško NPP site.

III-1. GROUND MOTION DATABASE

The local ground motion recordings in the model's database were provided by the Slovenian Environmental Agency [III–2]. The data included ground motion recordings from 1996 to 2020 for earthquakes with M≥2. Because the PSHA at the Krško NPP site is dominated by earthquakes at distances of about 25 km from the site, the ground motion recordings were selected by limiting the epicentral distances. In the case of low local magnitudes (M_L) (i.e. $3>M_L\geq 2$), it was decided to consider the earthquakes with epicentres located within 50 km from the Krško site, while for higher magnitudes (i.e. $M_L\geq 3$), earthquakes with epicentres located within 100 km of the Krško site were considered. For both sets, recordings out to an epicentral distance of 50 km were included.

Then, a subset of recordings was selected. Recordings from a given station were included in the subset if their number was at least four. Similarly, recordings from a given earthquake were used if four or more of such recordings were available. Only recordings from reference rock conditions were taken into account. The selected subset consisted of a total of 1078 recordings from 130 earthquakes. The epicentres of the considered earthquakes, the recording stations, the Krško site and the contour of the seismic source zone used in the simplified PSHA calculations (B-12) are presented in Fig. III–1.

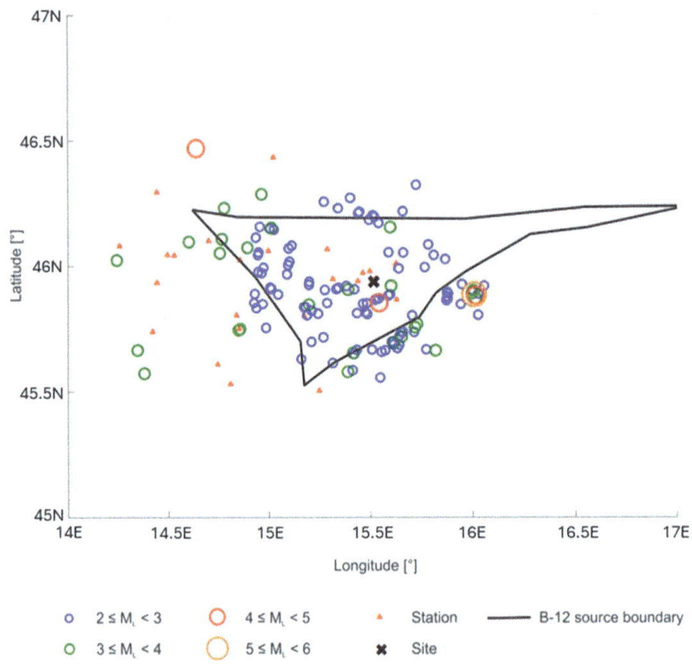

FIG. III–1. Earthquake epicentre locations with an indication of the local magnitude, recording station locations, the Krško site and the boundary of the seismic source zone B-12 (taken from [III–1]). Horizontal and vertical axes are the coordinates in longitude and latitude in degrees, respectively.

The ground motion database included velocity and acceleration time series. The velocity time series were first transformed into acceleration time series, and both recording types were converted to units of g. For each record, the Fourier amplitude spectra of the two horizontal components were computed using the full waveform. Next, the Fourier amplitude spectra for the two horizontal components were smoothed over frequency and combined to compute the effective amplitude spectra (EAS), which is consistent with inputs to random vibration theory to convert the Fourier amplitude spectral values to response spectral values. The EAS is the ground motion metric used by Ref. [III–3], which was used as the reference ergodic model to extrapolate the small magnitude data from Slovenia to larger magnitudes.

For small magnitude events, there can be an issue with the reliability of the moment magnitudes (M_W). The local magnitudes of the earthquakes in the Slovenia ground motion data set were converted to the moment magnitude, which is also used in the reference ergodic GMM [III–3]. The M_L-M_W conversion relation for $M_L \geq 3$ earthquakes was consistent with the M_L-M_W conversion used for the seismic source categorization model developed for the Krško site. Because this seismic source categorization model was developed only based on the earthquakes with $M_L \geq 3$, the conversion for $2 \leq M_L < 3$ earthquakes was evaluated by computing the 1-Hz EAS residuals relative to the reference ergodic model. For these small magnitude events, the 1-Hz EAS is below the corner frequency, so the residuals reflect errors in the M_L-M_W conversion and do not show differences in the stress drops. The average residuals over all the small magnitude events showed a bias of about 0.2 magnitude units. This bias was used to adjust the M_L-M_W relationship for $2 \leq M_L < 3$ earthquakes. The systematic differences in the ground motions between the Slovenia region and California that are the consequence of different median stress drops were maintained. These ground motion data were then used in the development of the non-ergodic GMM.

III-2. DEVELOPMENT OF THE NON-ERGODIC MODEL FOR FOURIER SPECTRA

In order to develop the non-ergodic GMM for Krško, the systematic source, site and path effects on the ground motions were first estimated in terms of frequency-dependent EAS. For this purpose, the following functional form was used:

$$EAS_{ln} = \left(GMM_{erg}(M,R,S) - c_7 R_{RUP}\right) + c_0 + \delta c_{1a}\left(x_{eq}\right) + \delta c_{1b}\left(x_{site}\right) + \sum_{i}^{Ncell} c_{7,i}\left(x_{cell,i}\right)\Delta R_i + \delta W_0 + \delta B_0 + \delta S2S \quad \text{(III–1)}$$

where EAS_{ln} is the EAS expressed in ln units, $GMM_{erg}(M,R,S)$ is the ergodic backbone GMM, c_7 is the coefficient in the ergodic GMM related to the linear source-to-site distance scaling for rupture distance, R_{RUP}, $\delta c_{1a}\left(x_{eq}\right)$ is the regional difference in the mean source term for earthquakes at coordinates x_{eq}, $\delta c_{1b}\left(x_{site}\right)$ is the regional difference in the mean site term for sites at coordinates x_{site}, $c_{7,i}\left(x_{cell,i}\right)$ is the linear distance scaling coefficient for the ray path in the ith cell, ΔR_i is the length of the ray path from the source to the site in the ith cell, δW_0 is the within-event aleatory variability, δB_0 is the between-event aleatory variability, $\delta S2S$ is the site term (uncorrelated between sites).

The linear distance term was removed from the ergodic GMM because it was replaced by a regionalized linear distance term, while the source and site terms of the non-ergodic model are presented by regional differences which are dependent on the earthquake and site coordinates, respectively. Note also that $\delta c_{1a}\left(x_{eq}\right)$, $\delta c_{1b}\left(x_{site}\right)$ and $c_{7,i}\left(x_{cell,i}\right)$ terms, which describe the systematic source, site and path effects, respectively, are not functions of the magnitude or source-to-site distance. On the contrary, they are functions of the frequency and were, therefore, estimated separately for each of the selected frequencies. Eleven frequencies were selected, ranging from 1 to 20 Hz. It was assumed that the evaluation of systematic effects at these frequencies is sufficient for further development of the non-ergodic GMM.

The main statistical tool to estimate the systematic source, site and path effects was the varying coefficient model, which uses Gaussian Process regression for the coefficient estimation, as explained in Section 6. This regression approach has two sets of terms: hyperparameters that describe the variance structure and non-ergodic terms that depend on location and describe the spatial changes in the median ground motion in the region. Because the local data from Slovenia were not sufficient for estimating the hyperparameters, the California model for the hyperparameters was adopted [III–4]. The hyperparameters consist of frequency-dependent correlation length (in km) and the corresponding standard deviation of δc_{1a}, δc_{1b} and $c_{7,i}$ from Eq. (III–1).

The results of the Gaussian Process regression are the mean coefficients from Eq. (III–1) and the corresponding epistemic uncertainty. Some of these coefficients are location-dependent (see Eq. (III–1)). The mean values and the corresponding standard deviations of the non-ergodic source, site and path terms, at the frequency of 5 Hz, are presented in Figs III–2 to III–4, respectively. In these figures, the black line and dots indicate, respectively, the border of

Slovenia and the locations of the earthquakes. The blue triangles show the locations of the recording stations. The red dot indicates the location of the Krško site. The boundary of the B-12 seismic source is shown by the thick brown line.

Figure III–2 shows that the source term does not vary significantly with the location. The mean values of δc_{1a} are mostly below 0.05. However, it can be concluded that the non-ergodic GMM predicts slightly higher ground motion intensities for an earthquake with a given magnitude if its epicentre is in the southeast of the map compared an earthquake with the epicentre in the northwest part. The epistemic uncertainty of the source term δc_{1a} is quite low, too.

The site term, δc_{1b}, exhibits significantly larger regional variability than the source term (Fig. III–3). In the vicinity of the Krško site, the standard deviation in the site term amounts to approximately 0.2, but the uncertainty in the site effect increases with the distance from the recording stations.

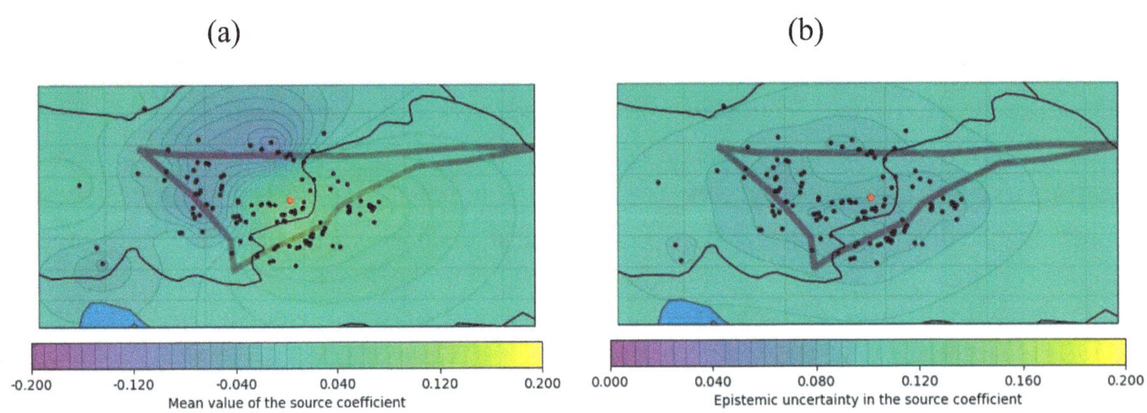

FIG. III–2. (a) The mean values of δc_{1a} for the frequency of 5 Hz and (b) the corresponding epistemic uncertainty in terms of standard deviation (taken from [III–1]).

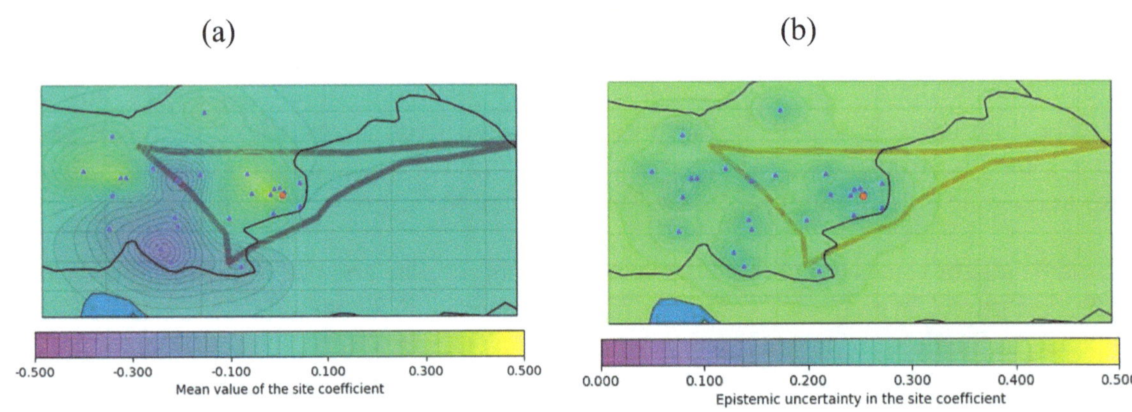

FIG. III–3. (a) The mean values of δc_{1b} for the frequency of 5 Hz and (b) the corresponding epistemic uncertainty in terms of standard deviation (taken from [III–1]).

Because the non-ergodic path term is the sum of partial path terms from each cell, the impact of the path term is presented by the mean difference in the anelastic attenuation coefficient between the non-ergodic and ergodic values ($c_{7,i} - c_7$) in the area close to the Krško site (Fig. III–4). At the frequency of 5 Hz, the most significant effect of the path term is in the regions northwest of the Krško site, where the mean ($c_{7,i} - c_7$) is negative. This suggests stronger attenuation of the ground motion as the seismic waves propagate through this specific area. The prominent effects occur at greater distances since this term is influenced by the linear distance scaling. For instance, if there is a variation of -0.01 in the non-ergodic anelastic attenuation, the ground motion at a distance of 20 km will be reduced by a factor of 1.2. However, at a distance of 50 km, the reduction is a factor of 1.6. Consequently, the contribution to the hazard from more distant fault sources in the area northwest from the Krško site will be diminished. The epistemic uncertainty in the anelastic attenuation at 5 Hz is low due to the extensive coverage of ray paths.

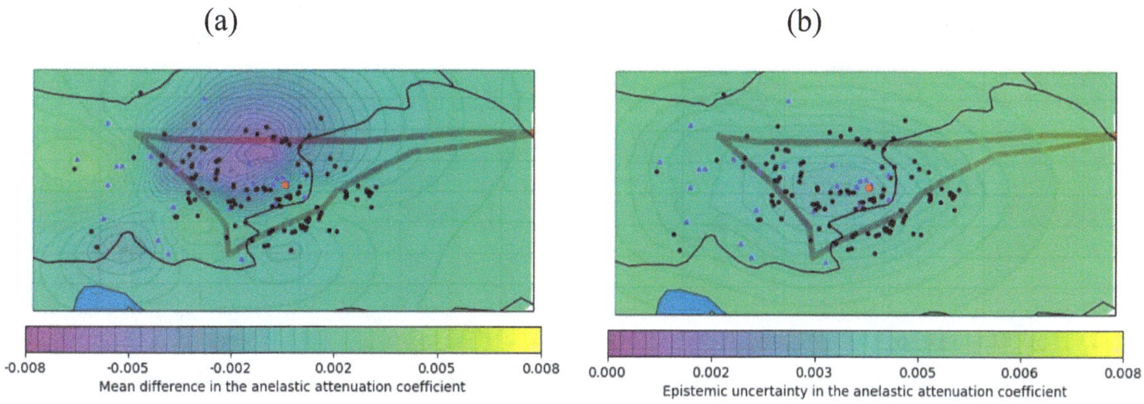

FIG. III–4. (a) The mean difference in the anelastic attenuation coefficient and (b) the corresponding epistemic uncertainty between the non-ergodic and ergodic values ($c_{7,i} - c_7$) in the area close to the Krško site for the frequency of 5 Hz (taken from [III–1]).

The mean non-ergodic adjustment, defined as the difference between the non-ergodic prediction of EAS_{ln} and the prediction made by the reference ergodic model, is presented in Fig. III–5 for the frequency of 5 Hz, at the Krško site and varying source locations. It can be concluded that earthquakes located northwest of the site have the strongest reduction, whereas earthquakes located close to the location of Krško have only a slight increase. The epistemic uncertainty in the total adjustment for this site is controlled by the epistemic uncertainty in the site term and is about 0.25 for epicentres at the vicinity of the Krško site and increases with the distance from the site. The non-ergodic aleatory standard deviation was calculated for small magnitude earthquakes from the dataset based on non-ergodic residuals using random effects for the earthquake term and site term. The value of 0.53 (in ln units) was obtained at 5 Hz, which is similar to 0.59, that was obtained for France in Ref. [III–5]. The estimated non-ergodic aleatory standard deviation for EAS_{ln} was used to estimate the non-ergodic aleatory standard deviation for spectral accelerations.

III-3. THE NON-ERGODIC MODEL FOR SPECTRAL ACCELERATIONS

The non-ergodic GMM for EAS_{ln} can be considered a stochastic model, which has to be realized by a number of samples for a straightforward implementation of the model into the PSHA. To enable the use of the non-ergodic model in engineering applications, samples of the non-ergodic model for EAS_{ln} have to be converted to samples of the non-ergodic spectral accelerations. Finally, the model for the non-ergodic standard deviation of ln values of spectral accelerations, i.e. non-ergodic sigma for spectral accelerations, has to be developed. These three steps are briefly explained in this section.

(a) (b)

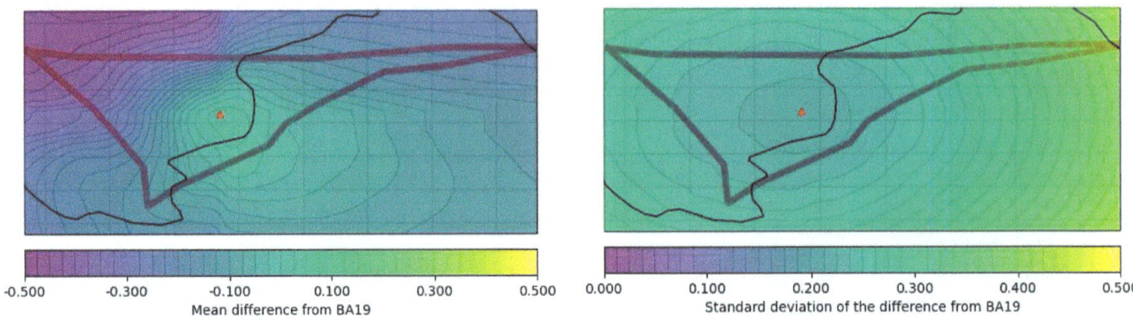

FIG. III–5. The (a) mean and (b) standard deviation of difference in the non-ergodic prediction of EAS_{ln} for the frequency of 5 Hz from the prediction made by the reference non-ergodic model for the Krško site and varying source locations (taken from [III–1]).

In the first step, 108 samples were defined, each consisting of the non-ergodic adjustments of EAS_{ln} for the eleven selected frequencies and 4154 source locations at a distance of approximately five km from each other. The samples were generated by considering the cross-correlation between different locations and different frequencies. The epistemic standard deviations and the correlations between different locations were obtained through the regression, while the correlations between different frequencies were modelled according to Ref. [III–4]. In the generation of samples, the epistemic standard deviation refers to only the parts related to the source, site, and path terms.

In the second step, the samples of the non-ergodic adjustments of EAS_{ln} were converted to the non-ergodic spectral acceleration adjustments, $\Delta\ln(SA)_{NGMM}$, which can be used in combination with an ergodic spectral acceleration backbone GMM to define a non-ergodic spectral acceleration GMM. The ergodic spectral acceleration backbone GMM needs to be consistent with the ergodic EAS backbone model used in the estimation of the non-ergodic adjustments of EAS_{ln}. In this study, the model proposed in Ref. [III–7] was selected as the spectral acceleration backbone GMM. The reference ergodic model for EAS and spectral acceleration backbone GMM are consistent because they were developed based on the same database and use the same functional form for magnitude scaling.

Each sample of the non-ergodic adjustments of EAS_{ln} was converted to a sample of non-ergodic spectral acceleration adjustments. Note that nine frequencies ranging from 1 to 100 Hz were selected for the spectral acceleration adjustments, which is different from the frequencies used in the estimation of the EAS_{ln} adjustments. This is not an issue because the spectral acceleration adjustment at a given frequency is not only a function of the EAS_{ln} adjustment at that same frequency but rather of the EAS_{ln} adjustments across a wider frequency range. Note also that while the EAS adjustment is not a function of the magnitude, the spectral acceleration adjustment is, because the spectral acceleration scale factors depend on the spectral shape. Therefore, the conversion of the EAS_{ln} adjustments to spectral acceleration adjustments was performed separately for moment magnitudes M_W=5, 6 and 7.

The conversion of EAS to spectral acceleration was done in the following steps:

1) The median EAS was computed using the reference ergodic model for EAS for the given magnitude and the distance corresponding to the given source location, assuming the kappa value as 0.035s for Slovenia using the random vibration theory [III–8, III–9].

2) The median EAS was combined with the given sample of the non-ergodic EAS_{ln} adjustments for the given source location. The adjusted EAS_{ln} was converted to spectral accelerations using the random vibration theory.
3) The ln values of spectral acceleration adjustment terms were calculated as the logarithms of the ratios between the spectral accelerations from Step 2 and spectral accelerations from Step 1 at the nine spectral acceleration related frequencies.

An example of the ln values of the mean non-ergodic spectral acceleration adjustments and the corresponding standard deviations are presented in Fig. III-6 for the frequency of 5 Hz and M_W=6.

(a) (b)

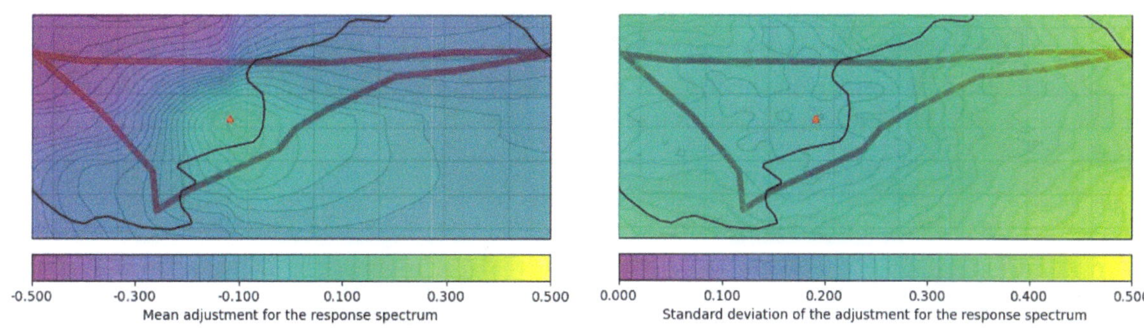

FIG. III–6. The mean non-ergodic ln values of spectral acceleration adjustments and the corresponding standard deviation for the frequency of 5 Hz, Mw=6, the Krško site and varying source location (taken from [III–1]).

For demonstrating the effect of the non-ergodic model on the PSHA, the 108 samples of the non-ergodic spectral acceleration adjustments were combined with the spectral acceleration backbone GMM. As a result, the epistemic uncertainty in the non-ergodic model for spectral accelerations was simulated by 108 logic tree branches, where each branch of the model was defined as follows:

$$\ln(SA)_{NGMM} = \ln\left(GMM_{erg}(M,R,S)\right) + \Delta\ln(SA)_{NGMM} \quad \text{(III–2)}$$

Note that Equation (III–2) is presented in the simplified form. As discussed above, the sample of 108 spectral acceleration adjustments in ln units were calculated for three magnitudes, for 11 frequencies and 4154 source locations.

Different approaches can be used to assess the non-ergodic sigma for spectral accelerations. For this case, the estimation of the non-ergodic sigma for spectral acceleration was based on the non-ergodic sigma for EAS, which was frequency-dependent and estimated based on small magnitude data. The procedure for estimating the non-ergodic sigma for spectral acceleration at the selected frequencies and for higher magnitudes was as follows:

1) For each frequency, the ratio between the ergodic standard deviation values from the reference ergodic EAS model corresponding to (i) small magnitude earthquakes, which are representative of the Slovenia dataset, and (ii) large magnitude earthquakes (M_W=5 or 6 or 7), which control the PSHA at the Krško site, were calculated. For the frequency of 5 Hz and M_W=6, this ratio was 1.18.
2) The non-ergodic sigma values for EAS estimated for small magnitude earthquakes using the Slovenia dataset were divided by the ratios calculated from the reference ergodic EAS model. For example, the non-ergodic sigma estimated with the Slovenia

dataset at the frequency of 5 Hz was 0.53. When divided by the ratio for M_W=6 estimated in Step 1, the non-ergodic sigma for EAS at M_W=6 was adjusted to 0.45.

3) The mean difference in the non-ergodic prediction of EAS_{ln} from the prediction made with the reference ergodic EAS model was shifted 100 times based on the non-ergodic standard deviations for EAS from Step 2 and the inter-frequency correlation model from [15]. This resulted in a sample of 100 maps of non-ergodic adjustments of EAS_{ln}. For each source location, the sample standard deviation was equal to the non-ergodic sigma from Step 2.
4) The sample of 100 non-ergodic adjustments of EAS_{ln} defined in Step 3 was converted to a sample of 100 non-ergodic spectral acceleration adjustments in ln units.
5) The standard deviation of spectral acceleration adjustments in ln units from Step 4 was calculated for each source locations and nine frequencies. The non-ergodic sigma for spectral acceleration at each frequency was then calculated as the average standard deviation for that frequency over all source locations within 30 km of the Krško site. The resulting value for spectral acceleration at 5 Hz was equal to 0.48, which is very similar to the value obtained by Ref. [III–4].

Described procedure was repeated for magnitudes 5, 6 and 7 that control the PSHA at the Krško site and used in PSHA calculations.

III-4. SIMPLIFIED HAZARD EXAMPLE

In the non-ergodic GMMs, the attenuation with distance can vary for different directions. As an example, using the backbone ergodic GMM for spectral accelerations, the distance attenuation for the ergodic GMM is compared to the direction-dependent distance attenuation with the non-ergodic adjustments in Fig. III–7. For the attenuation to the north, the non-ergodic GMM leads to a reduction in the T=0.2 s ground motion at a distance greater than 20 km. For the attenuation to the east, the non-ergodic GMM is similar to the attenuation for the ergodic GMM.

As an example of the effect of the non-ergodic GMM on the hazard, the 5 Hz spectral acceleration hazard based on the backbone ergodic GMM for spectral accelerations is compared to the hazard using the non-ergodic GMM in Fig. III–8. The non-ergodic GMM includes three changes from the reference ergodic GMM: (1) the change to the median ground motion for a given earthquake location, (2) the epistemic uncertainty in the non-ergodic term, and (3) the reduction in the aleatory standard deviation. Although there is only a small change in the median ground motion at short distances, as shown in Fig. III–7, the reduction in the aleatory standard deviation leads to steeper hazard curve and a reduction of the ground motion at long return periods as compared to the ergodic hazard.

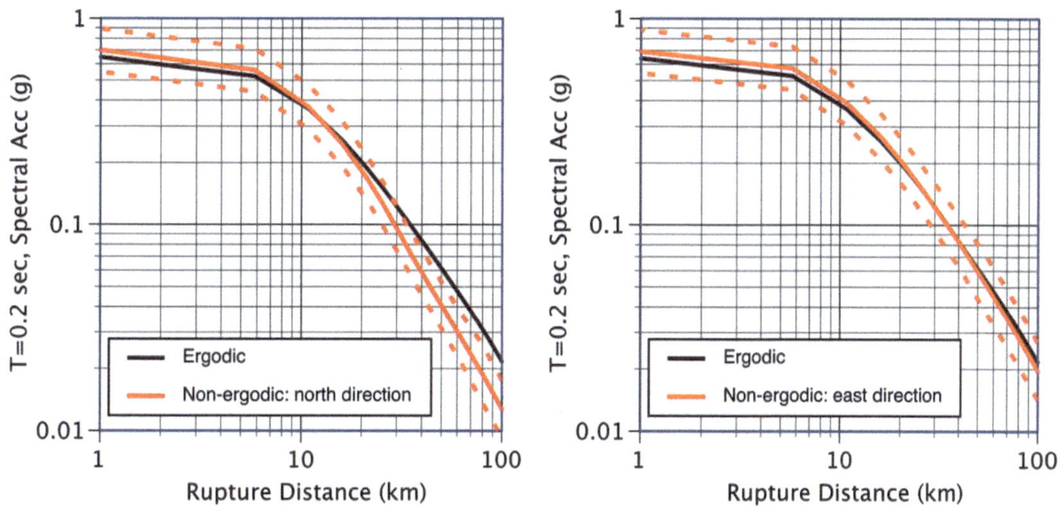

FIG. III–7. Comparison of the distance attenuation in the north and east directions for the non-ergodic GMM for 5 Hz (T=0.2 s) spectral acceleration for Mw=6, depth to top of rupture of 6 km, and V_{S30}=760 m/s. The dashed lines show ± 1 standard deviation of the non-ergodic terms. The epistemic uncertainty of the uncorrelated site terms (δS2S) is not included in the standard deviation of the non-ergodic terms because it will be included in the site response analysis.

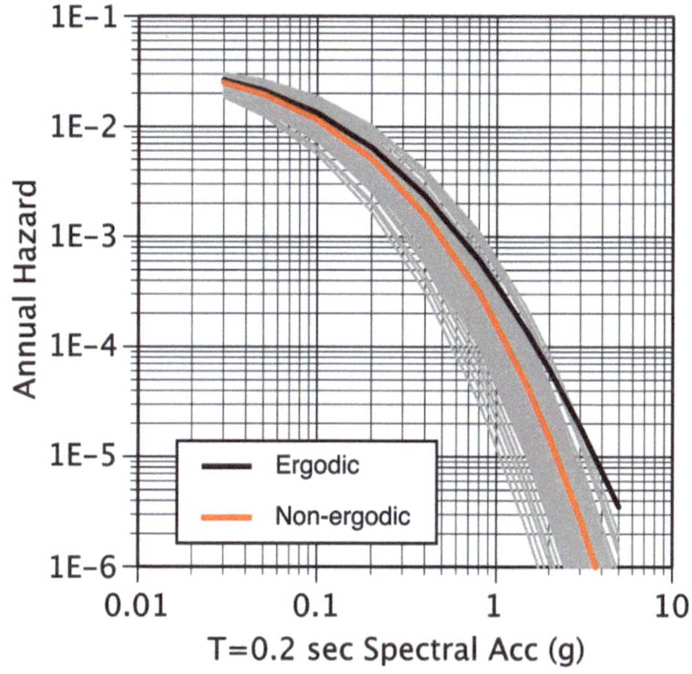

FIG. III–8. Comparison of the ergodic and non-ergodic hazard for 5 Hz (T=0.2 s) spectral acceleration. The grey curves show the hazard for the 108 realizations of the non-ergodic terms.

REFERENCES TO ANNEX III

[III–1] ABRAHAMSON, N., BABIČ, A., DOLŠEK, M., Task 1D – Non-ergodic Ground-Motion model for JEK2, Rev.2 (2024).

[III–2] ŽIVČIĆ, M., ČARMAN, M., ZUPANČIČ, P., LOŽAR STOPAR, M., Seismic records from ARSO waveform database, Slovenian Environment Agency (ARSO), Ljubljana, 2020.

[III–3] BAYLESS, J., ABRAHAMSON, N. A., Summary of the BA18 ground-motion model for Fourier amplitude spectra for crustal earthquakes in California, Bulletin of the Seismological Society of America **109** (2019) 2088-2105, https://doi.org/10.1785/0120190077.

[III–4] LAVRENTIADIS, G., ABRAHAMSON, N. A., KUEHN, N. M., A non-ergodic effective amplitude ground-motion model for California, Bulletin of Earthquake Engineering 21 (2023) 5233-5264, https://doi.org/10.1007/s10518-021-01206-w.

[III–5] SUNG, C-H. et al., A non-ergodic ground-motion model of Fourier amplitude spectra for France, Bulletin of Earthquake Engineering 21(11) (2023) 5293-5317, https://doi.org/10.1007/s10518-022-01403-1.

[III–6] BAYLESS, J., ABRAHAMSON, N.A., An empirical model for the inter-frequency correlation of epsilon for Fourier amplitude spectra, Bulletin of the Seismological Society of America **109** (2019) 1058-1070, https://doi.org/10.1785/0120180238.

[III–7] CHIOU, B., S.-J, YOUNGS, R. R., Update of the Chiou and Youngs NGA model for the average horizontal component of peak ground motion and response spectra, Earthquake Spectra **30** (2014) 1117-1153, https://doi.org/10.1193/072813EQS219M.

[III–8] BOORE, D.M. and JOYNER, W.B., A note on the use of random vibration theory to predict peak amplitudes of transient signals, Bulletin of the Seismological Society of America **74**(5) (1984) 2035-2039, https://doi.org/10.1785/BSSA0740052035.

[III–9] KOTTKE, A., RATHJE, E., Procedures for random vibration theory based seismic site response analyses, A White Paper Report Prepared for the Nuclear Regulatory Commission, Geotechnical Engineering Report GR08-09, Geotechnical Engineering Center, The University of Texas at Austin (2009).

LIST OF ABBREVIATIONS

EGF	empirical Green's function
FAS	Fourier amplitude spectra
GMICE	ground motion intensity conversion equations
GMM	ground motion model
GMPE	ground motion prediction equation
GP	Gaussian process
M_{MAX}	maximum magnitude
M_w	moment magnitude
NGA	next generation attenuation
NPP	nuclear power plant
PDF	probability distribution function
PGA	peak ground acceleration
PSA	Pseudo-spectral acceleration
PSHA	probabilistic seismic hazard assessment
R_{RUP}	closest distance of the site to the rupture plane or rupture distance
SCEC BBP	Southern California Earthquake Center Broadband Platform
SDOF	single degree of freedom system
SNR	signal to noise ratio
VCM	variable coefficient model
V_{S30}	time averaged shear wave velocity at the top 30 m of the soil profile
τ	between-event standard deviation
φ	within-event standard deviation

CONTRIBUTORS TO DRAFTING AND REVIEW

Abrahamson, N. A.	University of California, Berkeley, United States of America
Ameri, G.	SEISTER SAS, France
Babič, A.	Faculty of Civil and Geodetic Engineering, University of Ljubljana, Slovenia
Dolšek, M.	Faculty of Civil and Geodetic Engineering, University of Ljubljana, Slovenia
Fujiwara, H.	National Research Institute for Earth Science and Disaster Resilience, Japan
Fukushima, Y.	International Atomic Energy Agency
Gulerce, Z.	International Atomic Energy Agency
Graizer, V.	US Nuclear Regulatory Commission, United States of America
Hollender, F.	Commissariat à l'Energie Atomique, France
Kurmann, D.	Swissnuclear, Switzerland
Mamada, Y.	Nuclear Regulatory Authority, Japan
Malek, J.	Institute of Rock Structure and Mechanics of the Czech Academy of Sciences, Czech Republic
Marshall, N.R.	University of Oxford, United Kingdom
Nakajima, M.	Central Research Institute of Electric Power Industry, Japan
Nariyuki, M.	International Atomic Energy Agency
Prachar, I.	Consultant, Czech Republic
Renault, P.	SDA-engineering GmbH, Germany
Rodrigez-Marek, A.	Virginia Polytechnic Institute and State University, United States of America
Senfaute, Gloria	Electricité De France, France
Sugaya, K.	Nuclear Regulatory Authority, Japan
Sung, C-H.	University of California, Berkeley, United States of America
Traversa, P.	Electricité De France, France
Valentini, A.	International Atomic Energy Agency

Wu, C. Nuclear Regulatory Authority, Japan

Consultants Meetings

Vienna, Austria: 7–9 September 2021 (online), 9-10 May 2022, 6-8 December 2022, 29-30 August 2023

IAEA
International Atomic Energy Agency

CONTACT IAEA PUBLISHING

Feedback on IAEA publications may be given via the on-line form available at:
www.iaea.org/publications/feedback

This form may also be used to report safety issues or environmental queries concerning IAEA publications.

Alternatively, contact IAEA Publishing:

Publishing Section
International Atomic Energy Agency
Vienna International Centre, PO Box 100, 1400 Vienna, Austria
Telephone: +43 1 2600 22529 or 22530
Email: sales.publications@iaea.org
www.iaea.org/publications

Priced and unpriced IAEA publications may be ordered directly from the IAEA.

ORDERING LOCALLY

Priced IAEA publications may be purchased from regional distributors and from major local booksellers.

25-01625E

FSC
www.fsc.org
MIX
Papier | Fördert gute Waldnutzung
FSC® C083411

Zeitfracht Medien GmbH
Ferdinand-Jühlke-Straße 7
99095 Erfurt, Deutschland
produktsicherheit@kolibri360.de